SpringerBriefs in Mathematics

SpringerBriefs in Mathematics showcases expositions in all areas of mathematics and applied mathematics. Manuscripts presenting new results or a single new result in a classical field, new field, or an emerging topic, applications, or bridges between new results and already published works, are encouraged. The series is intended for mathematicians and applied mathematicians.

For further volumes:
http://www.springer.com/series/10030

Daniel Joseph Galiffa

On the Higher-Order Sheffer Orthogonal Polynomial Sequences

Daniel Joseph Galiffa
Penn State Erie
The Behrend College
Erie, Pennsylvania
USA

ISSN 2191-8198   ISSN 2191-8201 (electronic)
ISBN 978-1-4614-5968-2   ISBN 978-1-4614-5969-9 (eBook)
DOI 10.1007/978-1-4614-5969-9
Springer New York Heidelberg Dordrecht London

Library of Congress Control Number: 2012951348

Mathematics Subject Classification (2010): 33C45, 33C47

Printed on acid-free paper

Springer is part of Springer Science+Business Media (www.springer.com)

This monograph is dedicated to my nephew, Shadan Daniel

Preface

In 1939, I.M. Sheffer published seminal results regarding the characterizations of polynomials via general degree-lowering operators and showed that every polynomial sequence can be classified as belonging to exactly one *Type*. A large portion of his work was dedicated to developing a wealth of aesthetic results regarding the most basic type set, entitled *B-Type 0* (or equivalently *A-Type 0*), which included the development of several interesting characterizing theorems. In particular, one of Sheffer's most important results was his classification of which *B-Type 0* sets were also orthogonal, which are now known to be the very well-studied and applicable Laguerre, Hermite, Charlier, Meixner, Meixner–Pollaczek, and Krawtchouk polynomials, which are often called the Sheffer *Sequences*. As it turned out, Sheffer proved that every *B-Type 0* set $\{P_n(x)\}_{n=0}^{\infty}$ can be characterized by the generating function:

$$A(t)e^{xH(t)} = \sum_{n=0}^{\infty} P_n(x)t^n,$$

where $A(t)$ and $H(t)$ are formal power series in t, with certain restrictions. Furthermore, Sheffer also briefly described how this generating function can also be extended to the case of arbitrary *B-Type k* as follows:

$$A(t)\exp\left[xH_1(t) + \cdots + x^{k+1}H_{k+1}(t)\right] = \sum_{n=0}^{\infty} P_n(x)t^n,$$

$$\text{with } H_i(t) = h_{i,i}t^i + h_{i,i+1}t^{i+1} + \cdots, \quad h_{1,1} \neq 0, \quad i = 1, 2, \ldots, k+1.$$

Thus far, a very large amount of research has been completed regarding the theory and applications of the *B-Type 0* sets. Therefore, it is natural to attempt to determine whether or not orthogonal sets can be extracted from the higher-order classes, i.e., $k \geq 1$ in the generating relation directly above. In fact, no results have been published to date that *specifically* analyze the higher-order Sheffer classes. With this in mind, we have constructed the novel results of this monograph (Chap. 3), wherein we present a preliminary analysis of a special case of the *B-Type 1* class. We conduct this analysis for the following reasons. One, most importantly, our method functions as a template, which can be applied to other characterization problems as well. In order to effectively apply this method, computer algebra was found to be essential and Mathematica® was determined to be the most efficient platform for performing each of our manipulations. Therefore, our second motivation is the fact that the novel analysis of Chap. 3 lends itself as a paradigm regarding how computer algebra packages, like Mathematica®, can play an important role in developing rigorous mathematics. Lastly, we intend for this work to eventually lead to a complete characterization of the general *B-Type k* orthogonal sets and foster future research on other types of similar characterization problems as well.

We certainly wish to emphasize that Mathematica® was utilized only for *managing* many of the algebraic manipulations involved in establishing the original results within. The relationships achieved with the aid of Mathematica® were used to rigorously construct the novel theorems of this work, which were proven via algebraic techniques and rudimentary linear algebra, without the usage of computer algebra.

Now, it is well known that symbolic computations are becoming utilized more frequently in mathematical research and are also becoming increasingly more accepted. Several journals include various results that are based on computer algebra and some may even include, essentially, diminutive fragments of code, pseudo-code, or computer algebra outputs. In Chap. 3 of this work, a wealth of the Mathematica® inputs and their respective outputs are displayed in a reader-friendly format and written in a distinctive font class that is similar to the Mathematica® notebook. This amount of displayed code is certainly not typical in any of the current peer-reviewed mathematical science journals and is a luxury we are afforded in this monograph. Such uniquely detailed code displays are intended to increase the reader's understanding of our usage of Mathematica®, assist in verifications, and facilitate further experimentation.

It is also worth mentioning here that upon initially implementing the method of Chap. 3, it was also evident that the preliminary results were void of the level of elegance that mathematicians strive to achieve. Therefore, an approach was sought that would simultaneously give insights into existence/nonexistence of the *B-Type 1* orthogonal polynomial sequences and also yield a tractable problem that would ultimately admit elegant results. These goals were accomplished using simplifying assumptions that reduced the problem to a manageable format that was as similar in structure as possible to the *B-Type 0* class that Sheffer analyzed.

To enhance the novel results of this monograph, in Chap. 1 we additionally include an overview of the research that motivated their establishment. We begin by addressing Sheffer's derivation of the *B-Type 0* generating function defined above, as well as the characterizations of the *B-Type 0* orthogonal sets. Since, in 1934, J. Meixner initially studied this generating function and determined which sets were also orthogonal using a different approach than Sheffer, we also cover the central details of Meixner's method and results. We then briefly allude to W.A. Al–Salams's extension of Meixner's analysis.

Additionally, we discuss that there were actually three classifications that Sheffer developed: *A-Type*, *B-Type* (our focus in Chap. 3), and *C-Type*. The discrepancies between these types will also be addressed. We then present a summary of the σ-*Type* classification developed by E.D. Rainville, which is an extension of the Sheffer *A-Type* classification. Altogether, the Sheffer Sequences, the notion of *Type*, and the relevant background material are elucidated in order to facilitate the transition into the latter material.

To enhance this monograph even further, in Chap. 2 we discuss several of the many applications that classical orthogonal polynomials satisfy, which include first- and second-order differential equations quantum mechanics, difference equations and numerical integration. We first develop each of our applications in a general context and then show the specific roles played particular *A-Type 0* orthogonal sets. Through covering each of these applications, we also develop additional fundamental terminologies, definitions, lemmas, theorems, etc. that are very important in the field of orthogonal polynomials and special functions in a broad context. In essence, Chaps. 1 and 2 are intended to be used as (1) a concise, but informative, reference for developing new results related to the *A-Type 0* orthogonal sets and classical orthogonal polynomials in general and (2) provide material for advanced undergraduate courses, or graduate courses, in pure and applied analysis.

For the benefit of the reader, each chapter is self-contained. In addition, with respect to space constraints, this entire monograph has been written with as much detail, rigor, and supplementation via informative concrete examples as feasible.

This work represents the culmination of approximately three years of research on the Sheffer Sequences and related structures, which was conducted at Penn State Erie, The Behrend College.

Lastly, the author would like to thank Blair R. Tuttle for assisting in the proofreading of the overview of the Schrödinger equation in Chapter 2 and Jennifer K. Ulrich for assisting with the additional proofreading of the Brief.

Erie, PA, USA   Daniel Joseph Galiffa

Contents

Chapter 1
The Sheffer A-Type 0 Orthogonal Polynomial Sequences and Related Results

In this chapter, we present a rigorous development of I. M. Sheffer's characterization of the *A-Type 0* orthogonal polynomial sequences. We first develop the results that led to the main theorem that characterizes the general *A-Type 0* polynomial sequences via a linear generating function. From there, we develop the additional theory that Sheffer utilized in order to determine which *A-Type 0* polynomial sequences are also orthogonal. We then address Sheffer's additional characterizations of *B-Type* and *C-Type*, as well as E.D. Rainville's *σ-Type* classification. Lastly, we cover J. Meixner's approach to the same characterization problem studied by Sheffer and then discuss an extension of Meixner's analysis by W.A. Al-Salam. Portions of the analysis addressed throughout this chapter are supplemented with informative concrete examples.

1.1 Preliminaries

Throughout this chapter, we make use of each of the following definitions, terminologies and notations.

Definition 1.1. We always assume that a **set** of polynomials $\{P_n(x)\}_{n=0}^{\infty}$ is such that each $P_n(x)$ has degree exactly n, which we write as $\deg(P_n(x)) = n$.

Definition 1.2. A set of polynomials $\{Q_n(x)\}_{n=0}^{\infty}$ is **monic** if $Q_n(x) - x^n$ is of degree at most $n-1$ or equivalently if the leading coefficient of each $Q_n(x)$ is unitary.

Definition 1.3. We shall define a **generating function** for a polynomial sequence $\{P_n(x)\}_{n=0}^{\infty}$ as follows:

$$\sum_{\Lambda} \zeta_n P_n(x) t^n = F(x,t),$$

with $\Lambda \subseteq \{0,1,2,\ldots\}$ and $\{\zeta_n\}_{n=0}^{\infty}$ a sequence in n that is independent of x and t. Moreover, we say that the function $F(x,t)$ **generates** the set $\{P_n(x)\}_{n=0}^{\infty}$.

D.J. Galiffa, *On the Higher-Order Sheffer Orthogonal Polynomial Sequences*,
SpringerBriefs in Mathematics, DOI 10.1007/978-1-4614-5969-9_1,
© Daniel J. Galiffa 2013

It is important to mention that a generating function need not converge, as in general, several relationships can be derived when $F(x,t)$ is divergent.

Definition 1.4. In this chapter, the term *orthogonal polynomials* refers to a set of polynomials $\{P_n(x)\}_{n=0}^{\infty}$ that satisfies one of the two weighted inner products below:

$$\text{Continuous}: \quad \langle P_m(x), P_n(x) \rangle = \int_{\Omega_1} P_m(x)P_n(x)w(x)dx = \alpha_n \delta_{m,n}, \tag{1.1}$$

$$\text{Discrete}: \quad \langle P_m(x), P_n(x) \rangle = \sum_{\Omega_2} P_m(x)P_n(x)w(x) = \beta_n \delta_{m,n}, \tag{1.2}$$

where $\Omega_1 \subseteq \mathbb{R}$, $\Omega_2 \subseteq \mathbb{W}$, $\delta_{m,n}$ denotes the Kronecker delta and $w(x) > 0$ is entitled the *weight function*.

For example, the Laguerre, Hermite, and Meixner–Pollaczek polynomials satisfy a continuous orthogonality relation of the form (1.1). On the other hand, the Charlier, Meixner, and Krawtchouk polynomials satisfy a discrete orthogonality relation of the form (1.2) (cf. [6]).

Now, it is well-known that a necessary and sufficient condition for a set of polynomials $\{P_n(x)\}_{n=0}^{\infty}$ to be orthogonal is that it satisfies a three-term recurrence relation (see [8]), which can be written in different (equivalent) forms. In particular, we utilize the following two forms in this chapter and adhere to the nomenclature used in [2].

Definition 1.5 (The Three-Term Recurrence Relations). It is a necessary and sufficient condition that an orthogonal polynomial sequence $\{P_n(x)\}_{n=0}^{\infty}$ satisfies an *unrestricted three-term recurrence relation* of the form

$$P_{n+1}(x) = (A_n x + B_n)P_n(x) - C_n P_{n-1}(x), \quad A_n A_{n-1} C_n > 0$$

where $P_{-1}(x) = 0$ and $P_0(x) = 1$. $\tag{1.3}$

If $Q_n(x)$ represents the monic form of $P_n(x)$, then it is a necessary and sufficient condition that $\{Q_n(x)\}_{n=0}^{\infty}$ satisfies the following *monic three-term recurrence relation*:

$$Q_{n+1}(x) = (x + b_n)Q_n(x) - c_n Q_{n-1}(x), \quad c_n > 0$$

where $Q_{-1}(x) = 0$ and $Q_0(x) = 1$. $\tag{1.4}$

We entitle the conditions $A_n A_{n-1} C_n > 0$ and $c_n > 0$ above *positivity conditions*.

Lastly, we mention that all of the power series in this chapter are *formal* power series, i.e., they may or may not converge. In [9], Sheffer used the symbol '$\cong$' to denote formal series. For simplicity, we will use the equal sign throughout our present work and it will be tacitly assumed that each power series is nonetheless formal.

1.2   Sheffer's Analysis of the Type 0 Polynomial Sequences

In this section, we discuss each of the theorems of I.M. Sheffer's work [9] that were necessary in characterizing all of the *Type 0* orthogonal sets. With respect to space constraints, we write each proof, and some examples as well, with as much detail as possible. To begin, we consider the very well-studied Appell polynomial sets $\{P_n(x)\}_{n=0}^{\infty}$, which are defined as

$$A(t)e^{xt} = \sum_{n=0}^{\infty} P_n(x)t^n, \quad A(t) = \sum_{n=0}^{\infty} a_n t^n, \quad a_0 = 1. \tag{1.5}$$

An example of an Appell set is $\{x^n/n!\}_{n=0}^{\infty}$, which is clear since

$$e^{xt} = \sum_{n=0}^{\infty} \frac{x^n}{n!} t^n.$$

Now, we differentiate (1.5) with respect to x. The left-hand side becomes

$$\frac{d}{dx}\left[A(t)e^{xt}\right] = tA(t)e^{xt} = \sum_{n=0}^{\infty} P_n(x)t^{n+1} = \sum_{n=1}^{\infty} P_{n-1}(x)t^n$$

and the right-hand side becomes

$$\sum_{n=1}^{\infty} P_n'(x)t^n.$$

Therefore, after comparing coefficients of t^n in the results above, we achieve the equivalent characterization of Appell sets

$$P_n'(x) = P_{n-1}(x), \quad n = 0, 1, 2, \ldots.$$

Next, we consider the set of Newton polynomials $\{N_n(x)\}_{n=0}^{\infty}$, which is *not* an Appell set:

$$N_0(x) := 1, \quad N_n(x) := \frac{x(x-1)\cdots(x-n+1)}{n!}, \quad n = 1, 2, \ldots.$$

For the difference operator defined by $\Delta f(x) := f(x+1) - f(x)$, it can be shown that

$$\Delta N_n(x) = N_n(x+1) - N_n(x) = N_{n-1}(x)$$

and

$$(1+t)^x = e^{x\ln(1+t)} = \sum_{n=0}^{\infty} N_n(x)t^n.$$

We observe that the operator Δ functions as d/dx does on the Appell polynomials and that the generating function above is in a *more general* form than Eq. (1.5), i.e., the t in the exponent of Eq. (1.5) is replaced by $H(t) = \ln(t+1)$. Due to this analysis, Sheffer was motivated to define a class of *difference polynomial sets* that satisfy

$$J[P_n(x)] = P_{n-1}(x), \quad n = 0, 1, 2, \ldots$$

with J a general degree-lowering operator.

Thus, we now continue with a result regarding such a general degree-lowering operator J, which is an essential structure in all of the *Type 0* analysis that follows.

Lemma 1.1. *Assume that J is a linear operator that acts on the set of monomials $\{x^n\}_{n=0}^{\infty}$ such that $\deg(J[x^n]) \leq n$. Then, J has the following structure:*

$$J[y(x)] = \sum_{n=0}^{\infty} L_n(x) \frac{d^n}{dx^n} y(x), \qquad (1.6)$$

which is valid for all polynomials $y(x)$, with $\deg(L_n(x)) \leq n$.

Proof. We first note that since J is assumed to be a linear operator that acts on the set of monomials $\{x^n\}_{n=0}^{\infty}$, it can act on any polynomial. Therefore, if we show that Eq. (1.6) holds for $y(x) = x^n$, we have proven the theorem. Using the fact that

$$\frac{d^k}{dx^k} x^n = n(n-1)(n-2)\cdots(n-k+1)x^{n-k}$$

we can then recursively define $L_n(x)$ by the following:

$$J[x^n] = \sum_{k=0}^{n} L_k(x)[n(n-1)(n-2)\cdots(n-k+1)x^{n-k}], \quad n = 0, 1, 2, \ldots. \qquad (1.7)$$

Since for each $n = 0, 1, 2, \ldots$ we assumed that $\deg(J[x^n]) \leq n$, we must require that $L_k(x)[n(n-1)(n-2)\cdots(n-k+1)x^{n-k}]$ be of degree at most n for $k = 0, 1, \ldots, n$. This will occur if and only if $\deg(L_k(x)) \leq k$, since for any polynomials $P_m(x)$ and $Q_n(x)$, $\deg(P_m(x)Q_n(x)) = \deg(P_m(x)) + \deg(Q_n(x))$. $\qquad \square$

In Lemma 1.1, we determined the structure that J must adhere to in order for $\deg(J[x^n]) \leq n$. Next, we determine the form that J must have in order for $\deg(J[x^n]) = n - 1$. As we shall see, this will amount to restrictions on $L_n(x)$ in Eq. (1.6). Also, in order to naturally generalize our degree-lowering operator J, we additionally require that $J[c] = 0$ for all constants c, analogous to $\frac{d}{dx}[c] = 0$.

Lemma 1.2. *Necessary and sufficient conditions for J as defined in Eq. (1.6) to exist such that $\deg(J[x^n]) = n - 1$ are as follows:*

$$L_0(x) = 0, \quad L_n(x) = l_{n,0} + l_{n,1}x + \cdots + l_{n,n-1}x^{n-1}, \quad n = 1, 2, \ldots \qquad (1.8)$$

and

$$\lambda_n := nl_{1,0} + n(n-1)l_{2,1} + \cdots + n!l_{n,n-1} \neq 0, \quad n = 1,2,\ldots. \tag{1.9}$$

Proof. ($\Rightarrow$) We initially assume that $J[1] = 0$ and $\deg(J[x^n]) = n-1$ for $n = 1,2,\ldots$ and that $L_n(x)$ takes on the form

$$L_n(x) = l_{n,0} + l_{n,1}x + \cdots + l_{n,n}x^n,$$

from which we show that Eq. (1.8) and Eq. (1.9) necessarily follow. We begin by finding the coefficients of x^n and x^{n-1} in Eq. (1.7). Namely, we analyze the summand in Eq. (1.7):

$$L_k(x)[n(n-1)(n-2)\cdots(n-k+1)x^{n-k}] \tag{1.10}$$

for $k = 0,1,2,\ldots,n$ and determine the leading coefficient in each case and subsequently add the results, thus obtaining the coefficient of x^n, which will be valid for $n = 0,1,2,\ldots$. We follow the same procedure for achieving the coefficient of x^{n-1}.

For $k = 0$, we observe that Eq. (1.10) becomes $L_0(x)x^n = l_{0,0}x^n$, which clearly has a leading coefficient of $l_{0,0}$. For $k = 1$, we see that Eq. (1.10) turns out to be $L_1(x)nx^{n-1} = (l_{1,0} + l_{1,1}x)nx^{n-1}$ and therefore, the leading coefficient is $nl_{1,1}$. With $k = 2$, Eq. (1.10) becomes $L_2(x)n(n-1)x^{n-2} = (l_{2,0} + l_{2,1}x + l_{2,2}x^2)n(n-1)x^{n-2}$, which yields $n(n-1)l_{2,2}$ as the leading coefficient. Continuing in this fashion, we realize that for $k = n$ the leading coefficient of Eq. (1.10) is $n!l_{n,n}$. So, the coefficient of x^n is

$$l_{0,0} + nl_{1,1} + n(n-1)l_{2,2} + \cdots + n!l_{n,n}, \quad n = 0,1,2,\ldots. \tag{1.11}$$

We next successively compare (1.11) against $J[x^n]$ for $n = 0,1,2,\ldots$. For $n = 0$, Eq. (1.11) becomes $l_{0,0}$ and it must be that $l_{0,0} = 0$, since $J[1] = 0$, and thus $L_0(x) = 0$. With $n = 1$, Eq. (1.11) turns out to be $l_{1,1}$, which must be equal to zero, as $J[x] = \text{const.}$ Continuing in this manner, it follows that $l_{j,j} = 0$ for $j = 0,1,2,\ldots$, thus establishing (1.8).

Then, using the same logic that was used to determine the coefficient of x^n, we achieve the coefficient of x^{n-1} in Eq. (1.10), which we call λ_n, to be

$$\lambda_n := nl_{1,0} + n(n-1)l_{2,1} + \cdots + n!l_{n,n-1}, \quad n = 0,1,2,\ldots.$$

Since we have already shown that the coefficient of x^n is zero, in order to have $\deg(J[x^n]) = n-1$ we must also require that $\lambda_n \neq 0$, thus proving the necessity of the statement.

($\Leftarrow$) From substituting Eq. (1.8) with the restriction (1.9) into Eq. (1.7), the sufficiency of the statement is immediate.

$\square$

Due to Lemma 1.2, we can now modify the structure of Eq. (1.6), since our primary concern is when $\deg(J[x^n]) = n - 1$. We have

$$J[y(x)] = \sum_{n=1}^{\infty} [l_{n,0} + l_{n,1}x + \cdots + l_{n,n-1}x^{n-1}] \frac{d^n}{dx^n} y(x), \quad \lambda_n \neq 0, \quad n = 1, 2, \ldots. \quad (1.12)$$

The summation above starts at 1 via $L_0(x) = 0$.

Next, given a set of polynomials $\mathcal{S} = \{P_n(x)\}_{n=0}^{\infty}$, we want to determine how many operators J exist, such that J transforms each polynomial $P_k(x) \in \mathcal{S}$ to the polynomial immediately preceding it in the sequence, i.e., to $P_{k-1}(x) \in \mathcal{S}$. As it turns out, there is exactly one such operator.

Theorem 1.1. *For a given polynomial set* $\mathcal{S} = \{P_n(x)\}_{n=0}^{\infty}$, *there exists a unique operator* J *such that*

$$J[P_n(x)] = P_{n-1}(x), \quad n = 1, 2, \ldots \quad (1.13)$$

with $J[P_0(x)] := 0$.

Proof. To show the existence and uniqueness of J, we substitute $y(x) = P_n(x) \in \mathcal{S}$ into Eq. (1.12), which yields

$$J[P_n(x)] = \sum_{k=1}^{n} [l_{k,0} + l_{k,1}x + \cdots + l_{k,k-1}x^{k-1}] \frac{d^k}{dx^k} P_n(x).$$

Moreover, from Eq. (1.13), we require

$$\sum_{k=1}^{n} [l_{k,0} + l_{k,1}x + \cdots + l_{k,k-1}x^{k-1}] \frac{d^k}{dx^k} P_n(x) = P_{n-1}(x). \quad (1.14)$$

Therefore, upon successively comparing the coefficients of Eq. (1.14) for $n = 1, 2, \ldots$ we see that the sequence $\{l_{i,j}\}$ is uniquely determined, thus establishing the uniqueness of J given $\mathcal{S}$. $\qquad \square$

We say that the set $\{P_n(x)\}_{n=0}^{\infty}$ *corresponds* to the operator J if Eq. (1.13) is satisfied.

Example 1.1. To concretely demonstrate how the sequence $\{l_{i,j}\}$ is uniquely constructed, we consider J as in Eq. (1.12) acting on the Appell set

$$\mathcal{S} = \{x^n/n!\}_{n=0}^{\infty}.$$

First, for $n = 0$, we see that $J[1] = 0$ gives us no information. For $n = 1$, we require $J[x] = 1$. Therefore,

$$J[x] = \sum_{k=1}^{1} [l_{k,0} + l_{k,1}x + \cdots + l_{k,k-1}x^{k-1}] \frac{d^k}{dx^k} [x] = l_{1,0} \cdot 1,$$

which implies that $l_{1,0} = 1$. Then, for $n = 2$, we must have $J[x^2/2!] = x$ and thus

$$J\left[\frac{x^2}{2!}\right] = \sum_{k=1}^{2}[l_{k,0} + l_{k,1}x + \cdots + l_{k,k-1}x^{k-1}]\frac{d^k}{dx^k}\left[\frac{x^2}{2!}\right]$$

$$= l_{1,0} \cdot x + (l_{2,0} + l_{2,1}x) \cdot 1$$

$$= l_{2,0} + (1 + l_{2,1})x.$$

Therefore, $l_{2,0} = l_{2,1} = 0$. Next, for $n = 3$, we see that $J[x^3/3!] = x^2/2!$ must hold. Hence,

$$J\left[\frac{x^3}{3!}\right] = \sum_{k=1}^{3}[l_{k,0} + l_{k,1}x + \cdots + l_{k,k-1}x^{k-1}]\frac{d^k}{dx^k}\left[\frac{x^3}{3!}\right]$$

$$= l_{1,0} \cdot \frac{x^2}{2!} + (l_{2,0} + l_{2,1}x) \cdot x + (l_{3,0} + l_{3,1}x + l_{3,2}x^2) \cdot 1$$

$$= l_{3,0} + l_{3,1}x + \left(\frac{1}{2!} + l_{3,2}\right)x^2.$$

Therefore, it must be that $l_{3,0} = l_{3,1} = l_{3,2} = 0$.

In fact, continuing in this fashion, the interested reader can readily show that all of the l-values in this sequence will be uniquely determined to be zero, except $l_{1,0} = 1$. This is certainly clear since

$$\frac{d}{dx}\left[\frac{x^n}{n!}\right] = \frac{x^{n-1}}{(n-1)!}.$$

We next prove the converse of Theorem 1.1.

Theorem 1.2. *Associated to each operator J are infinitely many sets $\{P_n(x)\}_{n=0}^{\infty}$ such that Eq. (1.13) holds. More specifically, exactly one of these sets $\{B_n(x)\}_{n=0}^{\infty}$, entitled the* **basic set***, is such that*

$$B_0(x) = 1 \quad and \quad B_n(0) = 0, \quad n = 0, 1, 2, \ldots.$$

Proof. Let $Q_m(x)$ be a polynomial such that $\deg(Q_m) = m$. Then, via Eq. (1.12), we can construct a polynomial, say $P_{m+1}(x)$, such that $J[P_{m+1}(x)] = Q_m(x)$, where $\deg(P_{m+1}(x)) = m + 1$. However, since $J[\text{const}] = 0$, the polynomial $P_{m+1}(x)$ is unique only up to an additive constant. This proves the infinitude of sets $\{P_n(x)\}_{n=0}^{\infty}$ that correspond to a given J.

By assigning $B_0(x) := 1$ and assuming $B_n(0) = 0$ for $n > 0$, one can successively and uniquely determine the set $\{B_n(x)\}_{n=0}^{\infty}$ such that $J[B_n(x)] = B_{n-1}(x)$ and $\deg(B_n(x)) = n$.

$\square$

The next result states that for $\{P_n(x)\}_{n=0}^{\infty}$ to be a set corresponding to J, it must be expressed as a linear combination of polynomials from the basic set. However, the scalers in this linear combination appear in a special way and play a key role in the later characterizations.

Theorem 1.3. *A necessary and sufficient condition that $\{P_n(x)\}_{n=0}^{\infty}$ be a set corresponding to J is that there exist a sequence of constants $\{a_k\}$ such that*

$$P_n(x) = a_0 B_n(x) + a_1 B_{n-1}(x) + \cdots + a_n B_0(x), \qquad a_0 \neq 0. \qquad (1.15)$$

Proof. ($\Rightarrow$) Assume that $\{P_n(x)\}_{n=0}^{\infty}$ satisfies (1.15). Therefore, $\deg(P_n(x)) = n$ and by the linearity of J we have

$$J[P_n(x)] = J\left[\sum_{i=0}^{n} a_i B_{n-i}(x)\right] = \sum_{i=0}^{n} a_i J[B_{n-i}(x)] = \sum_{i=0}^{n-1} a_i B_{n-i-1}(x) = P_{n-1}(x),$$

which follows since $J[B_0(x)] = 0$ and $J[B_n(x)] = B_{n-1}(x)$ for $n > 0$ via Theorem 1.2. Thus $J[P_n(x)] = P_{n-1}(x)$.

($\Leftarrow$) We now assume that $\{P_n(x)\}_{n=0}^{\infty}$ corresponds to J. Since $\deg(B_n(x)) = n$, given $P_n(x)$, there must exist constants $\{a_{n,k}\}_{k=0}^{n}$ such that

$$P_n(x) = a_{n,0} B_n(x) + a_{n,1} B_{n-1}(x) + \cdots + a_{n,n} B_0(x), \qquad a_{n,0} \neq 0.$$

We act on this relation with J and see that the left-hand side becomes

$$J[P_n(x)] = P_{n-1}(x) = a_{n-1,0} B_{n-1}(x) + a_{n-1,1} B_{n-2}(x) + \cdots + a_{n-1,n-1} B_0(x)$$

and since $J[B_0(x)] = 0$, we see that the right-hand side turns into

$$a_{n,0} B_{n-1}(x) + a_{n,1} B_{n-2}(x) + \cdots + a_{n,n-1} B_0(x).$$

Therefore, from comparing coefficients of the results directly above, we infer that

$$a_{n,k} = a_{n-1,k}, \qquad k = 0, 1, \ldots, n-1.$$

Next, we momentarily fix k. Then, the relation immediately above implies that for all $n \geq k$ each $a_{n,k}$ is equal to $a_{n-1,k}$. Thus, the first index in the series $\{a_{n,k}\}$ is superfluous and thus can be omitted. We conclude that $\{a_k\}$ exist such that Eq. (1.15) is satisfied. □

It may at first appear counterintuitive that the elements of the sequence $\{a_k\}$ appear as they do in Eq. (1.15). However, the proof of Theorem 1.3 shows why this is the case. For emphasis, consider expressing a polynomial $P_n(x)$ corresponding to J as a linear combination of basic polynomials in the following "natural" way:

$$P_n(x) = a_n B_n(x) + a_{n-1} B_{n-1}(x) + \cdots + a_0 B_0(x).$$

Then,

$$J[P_n(x)] = J\left[\sum_{i=0}^{n} a_i B_i(x)\right] = \sum_{i=0}^{n} a_i J[B_i(x)] = \sum_{i=0}^{n} a_i B_{i-1}(x) \neq P_{n-1}(x).$$

We next wish to determine what conditions are needed for a set $\{Q_n(x)\}_{n=0}^{\infty}$ to correspond to J, given that a set $\{P_n(x)\}_{n=0}^{\infty}$ corresponds to J. As it turns out, $Q_n(x)$ must be written as a linear combination of $\{P_k(x)\}_{k=0}^{n}$.

Corollary 1.1. *Given that $\{P_n(x)\}_{n=0}^{\infty}$ is a set corresponding to J, a necessary and sufficient condition that $\{Q_n(x)\}_{n=0}^{\infty}$ also corresponds to J is that constants $\{b_k\}$ exist such that*

$$Q_n(x) = b_0 P_n(x) + b_1 P_{n-1}(x) + \cdots + b_n P_0(x), \qquad b_0 \neq 0.$$

Proof. The proof is similar to that of Theorem 1.3 and is left as an exercise for the reader. □

In light of the preceding theorems, we now state the definition of Sheffer *Type k*.

Definition 1.6. Let the set $\mathcal{S} := \{P_n(x)\}_{n=0}^{\infty}$ correspond to the unique operator J. Then, $\mathcal{S}$ is of **Sheffer Type k**, or simply **Type k**, if the coefficients $\{L_j(x)\}_{j=0}^{\infty}$ in Eq. (1.12) are such that $\deg(L_j(x)) \leq k$ for all j and there exists at least one $L_i(x) \in \{L_j(x)\}_{j=0}^{\infty}$ such that $\deg(L_i(x)) = k$. If $\{L_j(x)\}_{j=0}^{\infty}$ is unbounded, then $\mathcal{S}$ is of **Type ∞**.

With this definition, we have the following result.

Theorem 1.4. *There exist infinitely many sets for each Sheffer* Type k *(k finite or infinite).*

Proof. We know from Theorem 1.2 that associated to each operator J are infinitely many sets $\{P_n(x)\}_{n=0}^{\infty}$ such that Eq. (1.13) holds. This result is entirely independent of the degrees of the coefficients $\{L_j(x)\}_{j=0}^{\infty}$ in Eq. (1.12) and therefore the *Type*. Hence, Theorem 1.2 holds for all $\{L_j(x)\}_{j=0}^{\infty}$, even if it is unbounded, so there are infinitely many sets of every *Type* (finite or infinite). □

We now consider what effect replacing $\mathcal{S} := \{P_n(x)\}_{n=0}^{\infty}$ with $\{c_n P_n(x)\}_{n=0}^{\infty}$ has on the *Type* classification of $\mathcal{S}$. Assuming that $\mathcal{S}$ corresponds to J, we immediately observe that

$$J[c_n P_n(x)] = c_n J[P_n(x)] = c_n P_{n-1}(x) \neq c_{n-1} P_{n-1}(x).$$

This simple manipulation tells us that the *Type* is not necessarily preserved since we may need a new operator, say $\breve{J}$, such that $\breve{J}[c_n P_n(x)] = c_{n-1}P_{n-1}(x)$. We demonstrate this concretely in the following examples.

Example 1.2. In Example 1.1, we analyzed the Appell set $\mathcal{S} := \{x^n/n!\}_{n=0}^{\infty}$ and showed that $l_{1,0} = 1$ and that every other l-value was zero by utilizing (1.12). Moreover, we actually showed that $L_1(x) = 1$ and $L_j(x) \equiv 0$ for $j = 2, 3, \ldots$ and thus, that $\mathcal{S}$ is a *Type 0* set, since $k = 0$ in Definition 1.6. We next consider the set $\breve{\mathcal{S}} := \{c_n x^n/n!\}_{n=0}^{\infty}$, where $c_i \neq 0$ and each c_i is distinct.

For $n = 1$, we require $J[c_1 x] = c_0$. Therefore,

$$J[c_1 x] = \sum_{k=1}^{1}[l_{k,0} + l_{k,1}x + \cdots + l_{k,k-1}x^{k-1}]\frac{d^k}{dx^k}[c_1 x] = l_{1,0} \cdot c_1,$$

which implies that $l_{1,0} = c_0/c_1$ and therefore $L_1(x) = c_0/c_1$. Then, for $n = 2$, we must have $J[c_2 x^2/2!] = c_1 x$ and thus

$$J\left[\frac{c_2 x^2}{2!}\right] = \sum_{k=1}^{2}[l_{k,0} + l_{k,1}x + \cdots + l_{k,k-1}x^{k-1}]\frac{d^k}{dx^k}\left[\frac{c_2 x^2}{2!}\right]$$

$$= l_{1,0} \cdot c_2 x + (l_{2,0} + l_{2,1}x) \cdot c_2$$

$$= l_{2,0}c_2 + (c_0 c_2/c_1 + l_{2,1}c_2)x.$$

So, $l_{2,0} = 0$ and $l_{2,1} = (c_1^2 - c_0 c_2)/(c_1 c_2)$ giving $L_1(x) = \frac{(c_1^2 - c_0 c_2)}{c_1 c_2}x$. Therefore, we already see that $\breve{\mathcal{S}}$ is *not* a *Type 0* set. In fact, the interested reader can show that $\breve{\mathcal{S}}$ is actually *Type* ∞.

Example 1.3. We next consider a very important *Type 0* set, the importance of which will become most evident upon the completion of Sect. 1.3. This set is defined as $\mathcal{H}_n(x) := 2^{-n}H_n(x)/n!$, where

$$H_n(x) := 2^n n! \sum_{k=0}^{\lfloor n/2 \rfloor} \frac{(-1)^k x^{n-2k}}{2^{2k}k!(n-2k)!}$$

are the Hermite polynomials. For convenience, we write out the first four polynomials from the set $\{\mathcal{H}_n(x)\}_{n=0}^{\infty}$:

$$\mathcal{H}_0(x) = 1, \quad \mathcal{H}_1(x) = x, \quad \mathcal{H}_2(x) = \frac{1}{2}x^2 - \frac{1}{4} \quad \text{and} \quad \mathcal{H}_3(x) = \frac{1}{6}x^3 - \frac{1}{4}x.$$

We initially see that $J[\mathcal{H}_1(x)] = \mathcal{H}_0(x)$ implies

$$J[\mathcal{H}_1(x)] = \sum_{k=1}^{1}[l_{k,0} + l_{k,1}x + \cdots + l_{k,k-1}x^{k-1}]\frac{d^k}{dx^k}[\mathcal{H}_1(x)] = l_{1,0} = 1$$

and therefore, $L_1(x) = 1$. Then, $J[\mathcal{H}_2(x)] = \mathcal{H}_1(x)$ yields

$$J[\mathcal{H}_2(x)] = \sum_{k=1}^{2} [l_{k,0} + l_{k,1}x + \cdots + l_{k,k-1}x^{k-1}]\frac{d^k}{dx^k}[\mathcal{H}_2(x)]$$

$$= l_{2,0} + (1 + l_{2,1}x)x = x$$

and it must be that $l_{2,0} = l_{2,1} = 0$, i.e., $L_2(x) = 0$. Continuing, one concludes that $L_1(x) = 1$ and $L_j(x) = 0$ for $j = 2, 3, \ldots$ and thus $\{\mathcal{H}_n(x)\}_{n=0}^{\infty}$ is a *Type 0* set. Moreover, writing (1.12) in the operator form

$$J = \sum_{k=1}^{\infty} L_k(x)\frac{d^k}{dx^k}, \tag{1.16}$$

we find the unique operator J for our current set to be $J = d/dx$.

Example 1.4. Now consider the set $H_n(x) := H_n(x)/(n!)^2$, where $\{H_n(x)\}_{n=0}^{\infty}$ are the Hermite polynomials as defined in the last example. The first four polynomials from the set $\{H_n(x)\}_{n=0}^{\infty}$ are

$$H_0(x) = 1, \quad H_1(x) = 2x, \quad H_2(x) = x^2 - \frac{1}{2} \quad \text{and} \quad H_3(x) = \frac{2}{9}x^3 - \frac{1}{3}x.$$

Using the same procedure as in Examples 1.2 and 1.3, one can show that $L_1(x) = 1/2, L_2(x) = \frac{1}{2}x, L_3(x) = -1/4$, and $L_j(x) = 0$ for $j = 4, 5, \ldots$. Thus, $\{H_n(x)\}_{n=0}^{\infty}$ is a *Type 1* set and Eq. (1.16) becomes

$$J = \frac{1}{2}\frac{d}{dx} + \frac{1}{2}x\frac{d^2}{dx^2} - \frac{1}{4}\frac{d^3}{dx^3}.$$

Next, we notice that if $\{P_n(x)\}_{n=0}^{\infty}$ is a *Type 0* set, then each $L_n(x)$ must be a constant and we can therefore restate Definition 1.6 specifically for *Type 0* sets as follows.

Definition 1.7. $\{P_n(x)\}_{n=0}^{\infty}$ is a ***Type 0*** set if Eq. (1.13) holds with J defined by

$$J[y(x)] := \sum_{n=1}^{\infty} c_n y^{(n)}(x), \quad c_1 \neq 0. \tag{1.17}$$

We emphasize that as we have seen in Examples 1.2–1.4, Eq. (1.17) may or may not terminate, i.e., it may be finite *Type k* or *Type ∞*. We also have the following definition.

Definition 1.8. Let $J(t)$ be the formal power series

$$J(t) := \sum_{n=1}^{\infty} c_n t^n, \quad c_1 \neq 0,$$

which we entitle the **generating function for J**, with J as in Eq. (1.17).

Now, let the formal power series inverse of $J(t)$ be

$$H(t) := \sum_{n=1}^{\infty} s_n t^n, \quad s_1 = c_1^{-1} \neq 0. \tag{1.18}$$

This is a valid definition because if $J(t)$ is formally substituted for t in (1.18) and the coefficients are collected to form a single power series in t, then the coefficient of each t^n is a polynomial in $c_1, c_2, \ldots, c_n, s_1, s_2, \ldots, s_n$. Therefore, we can choose each s_n recursively and uniquely as a function of $c_1, c_2, \ldots, c_n, s_1, s_2, \ldots, s_{n-1}$ so the series has a single term t, i.e.,

$$J(H(t)) = H(J(t)) = t.$$

In considering $\exp(xH(t))$, we see that each coefficient of t^n in the formal power series expansion only comprises $s_1, s_2, \ldots, s_n$. Upon multiplying $\exp(xH(t))$ by

$$A(t) := \sum_{n=0}^{\infty} a_n t^n, \quad a_0 \neq 0,$$

we achieve a series in t where the coefficient of each t^n involves elements of the sequences $a_1, a_2, \ldots, a_n$ and $s_1, s_2, \ldots, s_n$, such that each coefficient of t^n is a polynomial in x of degree exactly n, adhering to Definition 1.1. This leads to the main result of this section.

Theorem 1.5. *The set $\{P_n(x)\}_{n=0}^{\infty}$ corresponds to the operator J and is of Sheffer Type 0 if and only if the sequence $\{a_n\}_{n=0}^{\infty}$ exists such that*

$$A(t)e^{xH(t)} = \sum_{n=0}^{\infty} P_n(x)t^n, \tag{1.19}$$

where

$$A(t) := \sum_{n=0}^{\infty} a_n t^n, \quad a_0 = 1 \quad \text{and} \quad H(t) := \sum_{n=1}^{\infty} s_n t^n, \quad s_1 = 1. \tag{1.20}$$

Proof. We show that both necessity and sufficiency will follow if we prove that the basic set $\{B_n(x)\}_{n=0}^{\infty}$ corresponding to J in Eq. (1.13) has the following generating function:

$$e^{xH(t)} = \sum_{n=0}^{\infty} B_n(x)t^n. \tag{1.21}$$

Since $\exp[xH(t)] = \sum_{n=0}^{\infty} [H^n(t)x^n/n!]$, we let this expansion have the form

$$e^{xH(t)} = \sum_{n=0}^{\infty} C_n(x)t^n. \tag{1.22}$$

Then, $\{C_n(x)\}_{n=0}^{\infty}$ must be such that $\deg(C_n(x)) = n$. We show that $\{C_n(x)\}_{n=0}^{\infty}$ is the basic set.

Upon setting $x = 0$ in Eq. (1.22) and comparing coefficients, we see immediately that $C_0(0) = 1$, and therefore $C_0(x) = 1$, and that $C_n(0) = 0$ for $n > 0$. Thus, $\{C_n(x)\}_{n=0}^{\infty}$ satisfies the initial conditions of the basic set. We clearly have $J[C_0(x)] = 0$ and next show that $J[C_n(x)] = C_{n-1}(x)$ for $n = 1, 2, 3, \ldots$. We apply J to Eq. (1.22):

$$J\left[\sum_{n=0}^{\infty} C_n(x)t^n\right] = \sum_{n=0}^{\infty} J[C_n(x)]t^n = J\left[e^{xH(t)}\right] = \sum_{n=0}^{\infty} c_n \frac{d^n}{dx^n}\left[e^{xH(t)}\right]$$

$$= c_1 \frac{d}{dx}\left[e^{xH(t)}\right] + c_2 \frac{d^2}{dx^2}\left[e^{xH(t)}\right] + c_3 \frac{d^3}{dx^3}\left[e^{xH(t)}\right] + \cdots$$

$$= \left(c_1 H(t) + c_2 H^2(t) + c_3 H^3(t) + \cdots\right) e^{xH(t)}$$

$$= J(H(t))e^{xH(t)} = te^{xH(t)} = \sum_{n=0}^{\infty} C_n(x)t^{n+1} = \sum_{n=0}^{\infty} C_{n-1}(x)t^n$$

(with $C_{-1}(x) := 0$), which follows since $H(t)$ and $J(t)$ are formal inverses of one another. By comparing coefficients of t^n in the relation directly above, we achieve $J[C_n(x)] = C_{n-1}(x)$ for $n = 1, 2, 3, \ldots$ and hence, $C_n(x) \equiv B_n(x)$ and Eq. (1.21) is established. Lastly, we now multiply Eq. (1.21) by $A(t)$ as in Eq. (1.20), which yields

$$A(t)e^{xH(t)} = \sum_{n=0}^{\infty} a_n t^n \sum_{n=0}^{\infty} B_n(x)t^n = \sum_{n=0}^{\infty} \sum_{k=0}^{n} a_k B_{n-k}(x)t^n = \sum_{n=0}^{\infty} P_n(x)t^n, \tag{1.23}$$

as a result of Theorem 1.3. $\qquad\square$

We also have the following.

Theorem 1.6. *The sequence of a_k-terms in Eq. (1.19) is exactly the same as those in Eq. (1.15).*

Proof. This statement follows immediately from the proof of Theorem 1.5, i.e., Eq. (1.23). $\qquad\square$

The next result plays an integral part in determining all of the Sheffer *Type 0* orthogonal polynomials in the next section. This characterization is also interesting unto itself, as it is expressed entirely in terms of elements of the generating function (1.19).

Corollary 1.2. $\{P_n(x)\}_{n=0}^{\infty}$ *is of Type 0 if and only if sequences $\{q_{k,0}\}$ and $\{q_{k,1}\}$ exist such that*

$$\sum_{k=1}^{\infty} (q_{k,0} + q_{k,1}x)P_{n-k}(x) = nP_n(x), \quad n = 1, 2, \ldots, \tag{1.24}$$

where

$$\left. \begin{array}{l} \sum_{k=0}^{\infty} q_{k+1,0}t^k = A'(t)/A(t) \\ \sum_{k=0}^{\infty} q_{k+1,1}t^k = H'(t) \end{array} \right\} \tag{1.25}$$

with $A(t)$ and $H(t)$ as defined in Eq. (1.20).

Proof. We assume that $\{P_n(x)\}_{n=0}^{\infty}$ is a *Type 0* set. We first consider the right-hand side of Eq. (1.24) and multiply it by t^n and sum for $n = 0, 1, 2, \ldots$. This yields

$$\sum_{n=0}^{\infty} nP_n(x)t^n = t \sum_{n=0}^{\infty} nP_n(x)t^{n-1} = t\frac{\mathrm{d}}{\mathrm{d}t}\left(A(t)e^{xH(t)}\right) = te^{xH(t)}\left(xA(t)H'(t) + A'(t)\right). \tag{1.26}$$

Since $J^k[P_n(x)] = P_{n-k}(x)$, we see that

$$\sum_{n=0}^{\infty} J^k[P_n(x)]t^n = t^k \sum_{n=k}^{\infty} P_{n-k}(x)t^{n-k} = t^k \sum_{m=0}^{\infty} P_m(x)t^m = t^k A(t)e^{xH(t)}.$$

Therefore, we now multiply the left-hand side of Eq. (1.24) by t^n, sum for $n = 0, 1, 2, \ldots$, and utilize (1.20) to obtain

$$\sum_{n=0}^{\infty}\sum_{k=1}^{\infty} (q_{k,0} + q_{k,1}x)P_{n-k}(x)t^n = \sum_{n=0}^{\infty}\sum_{k=1}^{\infty} (q_{k,0} + q_{k,1}x)J^k[P_n(x)]t^n$$

$$= \sum_{k=1}^{\infty} (q_{k,0} + q_{k,1}x) \sum_{n=0}^{\infty} J^k[P_n(x)]t^n = \sum_{k=1}^{\infty} (q_{k,0} + q_{k,1}x)t^k A(t)e^{xH(t)}$$

$$= \left(t\sum_{k=0}^{\infty} q_{k+1,0}t^k + tx\sum_{k=0}^{\infty} q_{k+1,1}t^k \right) A(t)e^{xH(t)} = te^{xH(t)}\left(xA(t)H'(t) + A'(t)\right). \tag{1.27}$$

Hence, we see that both Eqs. (1.26) and (1.27) are formally equal and from comparing coefficients of t^n we obtain Eq. (1.24). $\qquad\square$

1.3   The Type 0 Orthogonal Polynomials

We now determine which Sheffer *Type 0* sets are also orthogonal. We first assume that $\{Q_n(x)\}_{n=0}^{\infty}$ is an orthogonal set that satisfies the monic three-term relation of the form Eq. (1.4), which we write using the notation in [9]:

$$Q_n(x) = (x + \lambda_n)Q_{n-1}(x) + \mu_n Q_{n-2}(x), \quad \mu_n \neq 0, \quad n = 1, 2, \ldots. \tag{1.28}$$

In our first theorem of this section, we determine a necessary and sufficient form that the recursion coefficients λ_n and μ_n in Eq. (1.28) must have in order to characterize a *Type 0* orthogonal set. First, we note that since $\{Q_n(x)\}_{n=0}^{\infty}$ is an orthogonal set, then so is $\{c_n Q_n(x)\}_{n=0}^{\infty}$. Now, we want to determine for which sets $\{Q_n(x)\}_{n=0}^{\infty}$ satisfying (1.28) do there exist a sequence of constants $\{c_n\}_{n=0}^{\infty}$ such that

$$P_n(x) = c_n Q_n(x), \quad n = 0, 1, 2, \ldots \tag{1.29}$$

is a *Type 0* set. To accomplish this, we simultaneously use Eqs. (1.24) and (1.28), i.e., a characterization of a *Type 0* set and a characterization of an orthogonal set.

Theorem 1.7. *A necessary and sufficient condition that an orthogonal polynomial set* $\{Q_n(x)\}_{n=0}^{\infty}$ *satisfying* (1.28) *be such that* $P_n(x) = c_n Q_n(x)$ *is of* Type 0 *for some* $c_n \neq 0$ *is that* λ_n *and* μ_n *have the form*

$$\lambda_n = \alpha + bn \quad \text{and} \quad \mu_n = (n-1)(c+dn) \tag{1.30}$$

with $c + dn \neq 0$ *for* $n > 1$.

Proof. ($\Rightarrow$) We assume that Eq. (1.29) holds and show that λ_n and μ_n are as in Eq. (1.30). To begin, replace n with $n - k + 1$ in Eq. (1.28) and rewrite it in the following way:

$$xQ_{n-k}(x) = Q_{n-k+1}(x) - \lambda_{n-k+1}Q_{n-k}(x) - \mu_{n-k+1}Q_{n-k-1}(x). \tag{1.31}$$

Then, substituting the right-hand side of Eq. (1.29) into Eq. (1.24) and using Eq. (1.31), we see that

$$ncnQ_n(x) = \sum_{k=1}^{\infty} q_{k,0}c_{n-k}Q_{n-k}(x) + \sum_{k=1}^{\infty} q_{k,1}c_{n-k}[xQ_{n-k}(x)]$$

$$= \sum_{k=1}^{\infty} q_{k,0}c_{n-k}Q_{n-k}(x) + \sum_{k=1}^{\infty} q_{k,1}c_{n-k}Q_{n-k+1}(x)$$

$$- \sum_{k=1}^{\infty} q_{k,1}c_{n-k}\lambda_{n-k+1}Q_{n-k}(x) - \sum_{k=1}^{\infty} q_{k,1}c_{n-k}\mu_{n-k+1}Q_{n-k-1}(x). \tag{1.32}$$

We next compare coefficients of $Q_j(x)$ for $j = n, n-1, n-2$ using Eq. (1.32) to obtain relationships involving the q's and in turn develop expressions for the recursion coefficients λ_n and μ_n. First, by comparing the coefficients of $Q_n(x)$ we see that $nc_n = c_{n-1}q_{1,1}$ and iterating this relationship, we obtain

$$c_n = c_0 q_{1,1}^n / n!. \tag{1.33}$$

After comparing the coefficients of $Q_{n-1}(x)$, we realize that

$$0 = q_{1,0}c_{n-1} + q_{2,1}c_{n-2} - q_{1,1}c_{n-1}\lambda_n.$$

Dividing both sides of this equation by c_{n-2} and using Eq. (1.33) yield

$$\lambda_n = (q_{1,0}q_{1,1} + q_{2,1}(n-1))/q_{1,1}^2. \tag{1.34}$$

Lastly, upon comparing the coefficients of $Q_{n-2}(x)$, we achieve

$$0 = q_{2,0}c_{n-2} + q_{3,1}c_{n-3} - q_{2,1}c_{n-2}\lambda_{n-1} - q_{1,1}c_{n-1}\mu_n.$$

We also divide both sides of this equation by c_{n-2}, use Eq. (1.33), and also call upon Eq. (1.34) to obtain

$$\mu_n = (n-1)\left(q_{2,0}q_{1,1}^2 - q_{1,0}q_{1,1}q_{2,1} - (q_{1,1}^2 - q_{1,1}q_{3,1})(n-2)\right)/q_{1,1}^4. \tag{1.35}$$

Thus, we see that λ_n is at most linear in n and that μ_n is at most quadratic in n with a factor of $(n-1)$, i.e., λ_n and μ_n satisfy (1.30).

($\Leftarrow$) We assume that λ_n and μ_n agree with Eq. (1.30). We show that Eq. (1.29) is of *Type 0*. Now, if in Eq. (1.28) we replace $Q_n(x)$ with $c_n d_n Q_n(x) = d_n P_n(x)$ where $c_n := \alpha^n/n!$ $(\alpha \neq 0)$ and $d_n := c_n^{-1}$ then our three-term recurrence relation (1.28), with λ_n and μ_n as in Eq. (1.30), remains unaltered and we can therefore write

$$d_n P_n(x) = (x + \alpha + bn)d_{n-1}P_{n-1}(x) + (n-1)(c + dn)d_{n-2}P_{n-2}(x).$$

Dividing both sides of this relation by d_n defined above gives a relation of the form

$$nP_n(x) = (\alpha x + \beta + \gamma n)P_{n-1}(x) + (\delta + \varepsilon n)P_{n-2}(x),$$

where $\alpha \neq 0$, $\delta + \varepsilon n \neq 0$ for $n > 1$ and with $\beta = \alpha^2$, $\gamma = b\alpha$, $\delta = c\alpha^2$, and $\varepsilon = d\alpha^2$. This, of course, can be written in the form

$$xP_{n-1}(x) = \alpha^{-1}nP_n(x) - \alpha^{-1}(\beta + \gamma n)P_{n-1}(x) - \alpha^{-1}(\delta + \varepsilon n)P_{n-2}(x).$$

That is, we have expressed $xP_{n-1}(x)$ as a linear combination of $P_n(x)$, $P_{n-1}(x)$, and $P_{n-2}(x)$. Therefore, sequences $\{q_{k,0}\}$ and $\{q_{k,1}\}$ exist such that

$$T_n := (q_{1,0} + q_{1,1}x)P_{n-1}(x) + (q_{2,0} + q_{2,1}x)P_{n-2}(x) + \cdots = nP_n(x).$$

Furthermore, it can be shown that these q's are related in the following way:

$$q_{k+2,1} = \gamma q_{k+1,1} + \varepsilon q_{k,1}; \quad q_{1,1} = \alpha, \ q_{2,1} = \alpha\gamma, \tag{1.36}$$

$$q_{k+1,0} = \frac{1}{\alpha} \left(q_{k,1}(\delta - (k-1)\varepsilon) + q_{k+1,1}(\beta - \gamma k) + q_{k+2,1}(k+1) \right). \tag{1.37}$$

Thus, we have shown that $\{P_n(x)\}_{n=0}^{\infty}$ satisfies Corollary 1.2 and is therefore a *Type 0* set. □

As it turns out, we need the relations (1.36) and (1.37) to determine all of the *Type 0* orthogonal sets. Now, if Eq. (1.36) is known, then Eq. (1.37) can be determined. Therefore, we solve (1.36) by considering the characteristic equation

$$u^2 - \gamma u - \varepsilon = 0, \tag{1.38}$$

which has the solution

$$u_{1,2} = \left(\gamma \pm \sqrt{\gamma^2 + 4\varepsilon} \right)/2.$$

We analyze the different cases of the discriminant $\gamma^2 + 4\varepsilon$ and determine the value of $q_{k,1}$, and therefore $q_{k+1,0}$, in each case. Thus, in each case, this leads to expressions for $A(t)$ and $H(t)$ in Eq. (1.20) and therefore the generating function (1.19), which yields the corresponding orthogonal set. For the repeated root cases ($\gamma^2 + 4\varepsilon = 0$) we use the general solution structure $q_{k,1} = Au_1^k + Bku_1^k$ and for the distinct root cases ($\gamma^2 + 4\varepsilon \neq 0$), we use the general solution structure $q_{k,1} = Au_1^k + Bu_2^k$, where u_1 and u_2 are the roots listed above. There are a total of four cases, which we label as Ia, Ib, IIa and IIb. We work out the rigorous details of Case Ia and summarize the remaining cases.

Case Ia. $\gamma^2 + 4\varepsilon = 0$ and $\gamma \neq 0$
In this case, we see that Eq. (1.38) yields $u_1 = u_2 = \gamma/2$ as a solution, which implies

$$q_{k,1} = A(\gamma/2)^k + Bk(\gamma/2)^k.$$

Using our initial conditions in Eq. (1.36), we have the following system:

$$q_{1,1} = \frac{1}{2}A\gamma + \frac{1}{2}B\gamma = \alpha,$$

$$q_{2,1} = \frac{1}{4}A\gamma^2 + \frac{1}{4}B\gamma^2 = \alpha\gamma,$$

which has the solutions $A = 0$ and $B = 2\alpha/\gamma$ and thus,

$$q_{k,1} = k\alpha(\gamma/2)^{k-1}. \tag{1.39}$$

By substituting the right-hand side of Eq. (1.36) into Eq. (1.37) and then using Eq. (1.39) accordingly, after some algebraic manipulations, we obtain the following:

$$q_{k+1,0} = \left(\frac{\gamma}{2} \right)^{k-1} \left(\frac{1}{2}(\beta\gamma + 2\delta)k + \frac{1}{2}(\beta + \gamma)\gamma + \frac{1}{2}(\gamma^2 + 4\varepsilon)k \right).$$

Then, recalling that the discriminant is zero in this case, we obtain

$$q_{k+1,0} = \left(\frac{\gamma}{2}\right)^{k-1} \left(\frac{1}{2}(\beta\gamma + 2\delta)k + \frac{1}{2}(\beta + \gamma)\gamma\right). \tag{1.40}$$

For this case, we can now determine the series $H(t)$ and $A(t)$ as in Eq. (1.20) using Eq. (1.25). We first substitute (1.39) into our $H'(t)$ expression of Eq. (1.25) and then integrate, which leads to the geometric series

$$H(t) = \alpha t \sum_{k=0}^{\infty} \left(\frac{\gamma t}{2}\right)^k = \frac{2\alpha t}{2 - \gamma t}.$$

It is also worth mentioning that the formal inverse of $H(t)$ above is therefore readily determined to be

$$J(t) = \frac{2t}{2\alpha + \gamma t}.$$

For $A(t)$, we first notice that we have the following first-order differential equation:

$$A'(t) - \sum_{k=0}^{\infty} q_{k+1,0} t^k A(t) = 0,$$

which, via integrating factor, has the solution

$$A(t) = \mu \exp\left[\int \sum_{k=0}^{\infty} q_{k+1,0} t^k dt\right].$$

We then substitute (1.40) into the result directly above, evaluate the sum and integrate, which eventually yields

$$A(t) = \mu \left(\frac{2 - \gamma t}{2}\right)^{-2\delta(\gamma-2)/\gamma^2} \exp\left(\frac{2t(\beta\gamma + 2\delta)}{\gamma(2 - \gamma t)}\right),$$

with $(-2\delta(\gamma - 2)/\gamma^2)$ not equal to a nonnegative integer, so that $\mu_n \neq 0$ is satisfied.

Case Ib. $\gamma^2 + 4\varepsilon = 0$ and $\gamma = 0$
In this case, we also have $\varepsilon = 0$. This gives

$$q_{1,1} = \alpha, \quad q_{k,1} = 0 \quad (k > 1), \quad q_{1,0} = \beta, \quad q_{2,0} = \delta \quad \text{and} \quad q_{k,0} = 0 \quad (k > 2)$$

and thus,

$$H(t) = \alpha t, \quad J(t) = t/\alpha \quad \text{and} \quad A(t) = \mu \exp\left(\beta t + \frac{1}{2}\delta t^2\right),$$

with $\delta \neq 0$, so that $\mu_n \neq 0$ holds.

For Cases IIa and IIb that follow $(\gamma^2 + 4\varepsilon \neq 0)$, we first derive the general $q_{k,1}$ and $q_{k+1,0}$ terms. We have

$$q_{k,1} = \frac{\alpha(u_1^k - u_2^k)}{\sqrt{\gamma^2 + 4\varepsilon}} \quad \text{and} \quad q_{k+1,0} = \frac{\lambda u_1^k - \sigma u_2^k}{\sqrt{\gamma^2 + 4\varepsilon}} \tag{1.41}$$

with

$$\lambda := \delta + 2\varepsilon + (\beta + \gamma)u_1 \quad \text{and} \quad \sigma := \delta + 2\varepsilon + (\beta + \gamma)u_2.$$

Case IIa. $\gamma^2 + 4\varepsilon \neq 0$ and $\varepsilon = 0$
In this case, we consequently have $\gamma \neq 0$. Via Eq. (1.41), we observe that

$$H(t) = \alpha \int_0^t \frac{d\tau}{1 - \gamma\tau} = -\frac{\alpha}{\gamma}\ln(1 - \gamma t), \quad J(t) = \frac{1}{\gamma}\left(1 - \exp\left(-\frac{\gamma t}{\alpha}\right)\right),$$

$$A(t) = \mu(1 - \gamma t)^{-(\beta\gamma + \gamma^2 + \delta)/\gamma^2} \exp\left(-\frac{\delta t}{\gamma}\right),$$

with $\delta \neq 0$ so that $\mu_n \neq 0$.

Case IIb. $\gamma^2 + 4\varepsilon \neq 0$ and $\varepsilon \neq 0$
Here, from Eq. (1.41), one can achieve

$$H(t) = \alpha \int_0^t \frac{d\tau}{1 - \gamma\tau - \varepsilon\tau^2} = \alpha \frac{\ln((u_2 t - 1)/(u_1 t - 1))}{\sqrt{\gamma^2 + 4\varepsilon}},$$

$$J(t) = \frac{\exp\left(\frac{t}{\alpha}\sqrt{\gamma^2 + 4\varepsilon}\right) - 1}{u_1 \exp\left(\frac{t}{\alpha}\sqrt{\gamma^2 + 4\varepsilon}\right) - u_2}, \quad A(t) = \mu \frac{(1 - u_2 t)^{h_2}}{(1 - u_1 t)^{h_1}},$$

with

$$h_i = \frac{u_i(\beta + \gamma) + (\delta + 2\varepsilon)}{u_i\sqrt{\gamma^2 + 4\varepsilon}}, \quad i = 1, 2.$$

We next state the main result of this section and for simplicity, we redefine the parameters involved in each of our $H(t)$ and $A(t)$ expressions in the same manner as Sheffer. This result displays the *general forms* of each of the generating functions for the orthogonal *Type 0* sets.

Theorem 1.8. *A polynomial set* $\{P_n(x)\}_{n=0}^{\infty}$ *is* Type 0 *and orthogonal if and only if* $A(t)e^{xH(t)}$ *is of one of the following forms:*

$$A(t)e^{xH(t)} = \mu(1-bt)^c \exp\left(\frac{dt+atx}{1-bt}\right), \quad abc\mu \neq 0, \tag{1.42}$$

$$A(t)e^{xH(t)} = \mu \exp\left(t(b+ax)+ct^2\right), \quad ac\mu \neq 0, \tag{1.43}$$

$$A(t)e^{xH(t)} = \mu e^{ct}(1-bt)^{d+ax}, \quad abc\mu \neq 0, \tag{1.44}$$

$$A(t)e^{xH(t)} = \mu(1-t/c)^{d_1+x/a}(1-t/b)^{d_2-x/a}, \quad abc\mu \neq 0, \ b \neq c. \tag{1.45}$$

Proof. This statement follows from the above analysis. $\qquad\square$

By judiciously choosing each of the parameters in Eqs. (1.42)–(1.45) we can achieve all of the Sheffer *Type 0* orthogonal polynomials previously discussed. For emphasis, we write each of these parameter selections below and then display the corresponding generating function as it appears in contemporary literature.

The Laguerre Polynomials In Eq. (1.42), we select the parameters as $\mu = 1$, $a = -1$, $b = 1$, $c = -(\alpha+1)$, and $d = 0$ to obtain

$$\sum_{n=0}^{\infty} L_n^{(\alpha)}(x)t^n = (1-t)^{-(\alpha+1)} \exp\left(\frac{xt}{t-1}\right).$$

The Hermite Polynomials With the assignments $\mu = 1$, $a = 2$, $b = 0$, and $c = -1$ in Eq. (1.43), we have

$$\sum_{n=0}^{\infty} \frac{H_n(x)t^n}{n!} = \exp(2xt-t^2).$$

The Charlier Polynomials If in Eq. (1.44) we choose $\mu = 1$, $a = 1$, $b = 1/a$, $c = 1$, and $d = 0$, then we achieve

$$\sum_{n=0}^{\infty} \frac{C_n(x;a)t^n}{n!} = e^t\left(1-\frac{t}{a}\right)^x.$$

The Meixner Polynomials In Eq. (1.45), we select $\mu = 1$, $a = 1$, $b = 1$, $c =$ arbitrary constant, $d_1 = 0$, and $d_2 = -\beta$ leading to

$$\sum_{n=0}^{\infty} \frac{(\beta)_n}{n!} M(x;\beta,c)t^n = \left(1-\frac{t}{c}\right)^x (1-t)^{-(x+\beta)}.$$

The Meixner–Pollaczek Polynomials Taking $\mu = 1$, $a = -i$, $b = e^{i\phi}$, $c = e^{-i\phi}$, and $d_1 = d_2 = -\lambda$ in Eq. (1.45) leads to

$$\sum_{n=0}^{\infty} P_n^{(\lambda)}(x;\phi)t^n = (1-e^{i\phi}t)^{-\lambda+ix}(1-e^{-i\phi}t)^{-\lambda-ix}.$$

The Krawtchouk Polynomials Lastly, selecting $\mu = 1, a = 1, b = -1, c = p/(1 - p), d_1 = 0$, and $d_2 = N$ in Eq. (1.45) yields

$$\sum_{n=0}^{N} C(N,n)K_n(x; p,N)t^n = \left(1 - \frac{1-p}{p}t\right)^x (1+t)^{N-x},$$

for $x = 0, 1, 2, \ldots, N$, where $C(N,n)$ denotes the binomial coefficient.

Refer to [6] and the references therein for definitions and characterizations of each of these *A-Type 0* sets. For additional analyses regarding these orthogonal sets, also consider [3–5, 8, 10].

1.4 An Overview of the Classification of Type

We now discuss the essential details of the three kinds of characterizations that Sheffer developed in [9], which are entitled *A-Type*, *B-Type* and *C-Type*. The definition of *Type* is dependent on which characterization of *Type 0* that is to be generalized. That is, each of the Sheffer Types generalizes a certain *Type 0* characterizing structure.

1.4.1 The Sheffer A-Type Classification

We deem all of the characterizations up to this juncture as *A-Type k* and therefore restate Definition 1.6 accordingly.

Definition 1.9. Let the set $\mathcal{S} := \{P_n(x)\}_{n=0}^{\infty}$ correspond to the unique operator J. Then, $\mathcal{S}$ is of *A-Type k* if the coefficients $\{L_j(x)\}_{j=0}^{\infty}$, as defined in Eq. (1.12), are such that $\deg(L_j(x)) \leq k$ for all i and there exists at least one $L_i(x) \in \{L_j(x)\}_{j=0}^{\infty}$ such that $\deg(L_i(x)) = k$. If $\{L_j(x)\}_{j=0}^{\infty}$ is unbounded, then $\mathcal{S}$ is of *A-Type ∞*.

We also will make use of the following result.

Theorem 1.9. *The set* $\{P_n(x)\}_{n=0}^{\infty}$ *is of* A-Type 0 *if and only if a sequence of constants* $\{s_n\}$ *exist such that*

$$P_n'(x) = s_1 P_{n-1}(x) + s_2 P_{n-2}(x) + \cdots + s_n P_0(x), \qquad n = 1, 2, 3, \ldots, \qquad (1.46)$$

where the elements of the sequence $\{s_n\}$ *are the same as those in Eq.* (1.18).

Proof. We differentiate (1.19) with respect to x and see that

$$\sum_{n=0}^{\infty} P_n'(x)t^n = H(t)A(t)e^{xH(t)} = \sum_{n=0}^{\infty} P_n(x)H(t)t^n.$$

Then, from comparing coefficients of t^n in

$$\sum_{n=0}^{\infty} P_n'(x)t^n = \sum_{n=0}^{\infty} P_n(x)[s_1 t + s_2 t^2 + s_3 t^3 + \cdots]t^n,$$

we obtain our result. It is readily seen that this argument is both necessary and sufficient.   □

1.4.2 The Sheffer B-Type Classification

To begin, we state that Theorem 1.9 can be extended in a natural way. Consider the structure

$$T_0(x)P_{n-1}(x) + T_1(x)P_{n-2}(x) + \cdots + T_{n-1}(x)P_0(x), \quad n = 1,2,3,\ldots.$$

Then, by successively setting $n = 1,2,\ldots$, we observe that given a set $\{P_n(x)\}_{n=0}^{\infty}$, a unique sequence of polynomials $\{T_n(x)\}$ exists such that

$$P_n'(x) = T_0(x)P_{n-1}(x) + T_1(x)P_{n-2}(x) + \cdots + T_{n-1}(x)P_0(x), \quad n = 1,2,\ldots \quad (1.47)$$

with $T_0(x) \neq 0$. However, $\deg(T_n(x)) \leq n$ since, e.g., Eq. (1.47) readily reduces to Eq. (1.46) if $\{P_n(x)\}_{n=0}^{\infty}$ is of *A-Type 0*. The following statement immediately follows.

Theorem 1.10. *For every set $\{P_n(x)\}_{n=0}^{\infty}$ there exist a unique sequence of polynomials $\{T_n(x)\}$, with $\deg(T_n(x)) \leq n$, such that Eq. (1.47) holds.*

The converse of this theorem does not hold, as we have the following result.

Theorem 1.11. *Given a sequence of polynomials $\{T_n(x)\}$, with $\deg(T_n(x)) \leq n$, there exist infinitely many sets $\{P_n(x)\}_{n=0}^{\infty}$ such that Eq. (1.47) is satisfied.*

Proof. This result is immediate, since the constant term of $P_n(x)$ in Eq. (1.47) can be made arbitrary.   □

Based on the above results, we have the following definition.

Definition 1.10. A set $\{P_n(x)\}_{n=0}^{\infty}$ is of **B-Type k** if the maximum degree of the polynomials $\{T_n(x)\}$ in Eq. (1.47) is k. Otherwise, $\{P_n(x)\}_{n=0}^{\infty}$ is classified as **B-Type ∞**.

We can now establish the following statement.

Theorem 1.12. *A set $\{P_n(x)\}_{n=0}^{\infty}$ is of B-Type 0 if and only if it is of A-Type 0.*

Proof. ($\Rightarrow$) If $\{P_n(x)\}_{n=0}^{\infty}$ is of *B-Type 0*, then $\{T_n(x)\}$ in Eq. (1.47) must be a sequence of constants. Moreover, Eq. (1.47) is reduced to Eq. (1.46) and thus $\{P_n(x)\}_{n=0}^{\infty}$ is an *A-Type 0* set.
($\Leftarrow$) If $\{P_n(x)\}_{n=0}^{\infty}$ is of *A-Type 0*, then Eq. (1.46) holds, which is Eq. (1.47) with constant coefficients. Thus, $\{P_n(x)\}_{n=0}^{\infty}$ is of *B-Type 0*. $\square$

We also have the classification of *B-Type k* sets below.

Theorem 1.13. *The characterization* (1.47) *is equivalent to the relation*

$$H(x,t) = A(t)\exp\left[t\int_0^x T(\xi,t)\mathrm{d}\xi\right],\qquad(1.48)$$

with $A(t)$ as in Eq. (1.20).

Proof. ($\Rightarrow$) First, we define

$$H := H(x,t) = \sum_{k=0}^{\infty} P_k(x)t^k,\qquad(1.49)$$

$$T := T(x,t) = \sum_{k=0}^{\infty} T_k(x)t^k.\qquad(1.50)$$

We now consider

$$tHT = t\sum_{k=0}^{\infty} P_k(x)t^k \sum_{k=0}^{\infty} T_k(x)t^k.\qquad(1.51)$$

We can find the coefficient of t^n in the right-hand side of Eq. (1.51) by considering the following structure, which is t multiplied by the general term in each of the sums of Eq. (1.51):

$$tP_{k_0}t^{k_0}T_{k_1}t^{k_1}$$

and finding all nonnegative integer solutions to the equation $1 + k_0 + k_1 = n$, i.e., the sum of the t-exponents. For the solution $k_0 = n - 1$ and $k_1 = 0$, we obtain $T_0(x)P_{n-1}(x)$ and for the solution $k_0 = n - 2$ and $k_1 = 1$, we achieve $T_1(x)P_{n-2}(x)$. Continuing in this fashion, we see that the coefficient of t^n turns out to be

$$T_0(x)P_{n-1}(x) + T_1(x)P_{n-2}(x) + \cdots + T_{n-1}(x)P_0(x) = P_n'(x)$$

via Eq. (1.47). Thus, the right-hand side of Eq. (1.51) is $\sum_{n=0}^{\infty} P_n'(x)t^n = \partial H(x,t)/\partial x$ and we have constructed the following first-order differential equation, as Eq. (1.51) becomes

$$\frac{\partial}{\partial x}H - tTH = 0;\qquad H(0,t) = \sum_{n=0}^{\infty} P_n(0)t^n = A(t),\qquad(1.52)$$

with $A(t)$ as in Eq. (1.20). Furthermore, we have shown that this relation is equivalent to Eq. (1.47). The general solution to Eq. (1.52) can be determined via the integrating factor $\exp\left[-\int_0^x tT(\xi,t)\mathrm{d}\xi\right]$ to be

$$H(x,t) = A(t)\exp\left[t\int_0^x T(\xi,t)\mathrm{d}\xi\right],$$

where $A(t)$ is an arbitrary power series with a nonzero constant term. However, in considering the initial condition of Eq. (1.52), we see that $A(t)$ must be as in Eq. (1.20). This argument is certainly both necessary and sufficient. $\square$

Next, we note that if we write $T(x,t)$ in Eq. (1.50) as a power series in x, as opposed to t, the coefficients of each of the x^n-terms are power series in t. To facilitate this, momentarily let $T_n(x) = c_{n,n}x^n + c_{n,n-1}x^{n-1} + \cdots + c_{n,0}$, and then we have

$$T(x,t) = (c_{0,0} + c_{1,0}t + c_{2,0}t^2 + \cdots) + (c_{1,1}t + c_{2,1}t^2 + c_{3,1}t^3 + \cdots)x$$
$$+ (c_{2,2}t^2 + c_{3,2}t^3 + c_{4,2}t^4 + \cdots)x^2 + \cdots.$$

Letting $x = \xi$ above, integrating this result with respect to ξ from 0 to x and multiplying by t, we see that

$$t\int_0^x T(\xi,t)\mathrm{d}\xi = xH_1(t) + x^2 H_2(t) + x^3 H_3(t) + \cdots,$$

where each $H_i(t)$ is a power series in t that starts with the t^i-term (or higher if the coefficient of the t^i-term is zero), i.e., $H_i(t) = h_{i,i}t^i + h_{i,i+1}t^{i+1} + h_{i,i+2}t^{i+2} + \cdots$. However, $H_1(t)$ must begin with a linear term in t, since by Eq. (1.47) $T_0(x) \neq 0$ and $h_{1,1} = T_0(x)$ by construction. Thus, we can write (1.48) in the following way:

$$H(x,t) = A(t)\exp\left[xH_1(t) + x^2 H_2(t) + x^3 H_3(t) + \cdots\right], \quad h_{1,1} \neq 0. \qquad (1.53)$$

Now, for $\{P_n(x)\}_{n=0}^{\infty}$ to be of B-Type k, it is necessary and sufficient that $T(x,t)$ is a polynomial in x of degree k. This restriction is equivalent to terminating the sum $\left[xH_1(t) + x^2 H_2(t) + x^3 H_3(t) + \cdots\right]$ in Eq. (1.53) at $k+1$ via the presence of the integral in Eq. (1.48). Thus, we obtain (1.54) in Theorem 1.14 below, which is the culmination of the above analysis.

Theorem 1.14. *A set is of* B-Type k *if and only if*

$$H(x,t) = A(t)\exp\left[xH_1(t) + \cdots + x^{k+1}H_{k+1}(t)\right],$$

with $H_i(t) = h_{i,i}t^i + h_{i,i+1}t^{i+1} + \cdots, \quad h_{1,1} \neq 0, \quad i = 1, 2, \ldots, k+1.$ (1.54)

1.4.3 The Sheffer C-Type Classification

The last classification that Sheffer developed is entitled *C-Type*. As was done in the previous section, we establish all of the theorems necessary for understanding this classification, beginning with the following.

Theorem 1.15. *For each set* $\{P_n(x)\}_{n=0}^{\infty}$, *there exist a unique sequence of polynomials* $\{U_n(x)\}_{n=0}^{\infty}$, *with* $\deg(U_n(x)) \le n$, *such that*

$$nP_n(x) = U_1(x)P_{n-1}(x) + U_2(x)P_{n-2}(x) + \cdots + U_n(x)P_0(x), \quad n = 1, 2, \ldots. \quad (1.55)$$

Proof. Analogous to establishing (1.47) in Theorem 1.10, we set $n = 1, 2, \ldots$ in Eq. (1.55) and then successively and uniquely determine the set $\{U_n(x)\}_{n=0}^{\infty}$. Through this process, it becomes clear that no polynomial $U_n(x)$ can exceed the degree of its subscript. The result therefore follows. □

This leads us to the following definition.

Definition 1.11. A set $\{P_n(x)\}_{n=0}^{\infty}$ is classified as *C-Type* k if the maximum degree of the polynomials $\{U_n(x)\}$ in Eq. (1.55) is $k + 1$. Else, $\{P_n(x)\}_{n=0}^{\infty}$ is classified as *C-Type* ∞.

We note that we use $k + 1$ in the above definition, as opposed to k as in our previous *Type* definitions, since otherwise, Eq. (1.55) would not be satisfied. As an example, consider the case when each $U_i(x)$ is constant.

Now, we define

$$U := U(x, t) = \sum_{n=0}^{\infty} U_{n+1}(x) t^n.$$

Then, calling upon Eq. (1.55) and using the same methodology that was used to construct (1.52) in the previous section, we derive the following first-order differential equation:

$$\frac{\partial}{\partial t} H - UH = 0; \quad H(x, 0) = P_0(x) = a_0$$

with $H := H(x, t)$ as in Eq. (1.49). This can easily be solved using the integrating factor $\exp\left[-\int_0^t U(x, \tau) d\tau\right]$, yielding the solution

$$H(x, t) = a_0 \exp\left[\int_0^t U(x, \tau) d\tau\right], \quad (1.56)$$

after incorporating our initial condition. Then, from comparing (1.56) with Eq. (1.48), we have

$$a_0 \exp\left[\int_0^t U(x, \tau) d\tau\right] = A(t) \exp\left[t \int_0^x T(\xi, t) d\xi\right]$$

and taking the natural logarithm of both sides of this relation leads to

$$\ln a_0 + \int_0^t U(x,\tau)\mathrm{d}\tau = \ln A(t) + t\int_0^x T(\xi,t)\mathrm{d}\xi. \tag{1.57}$$

Next, differentiate (1.57) with respect to t in order to obtain

$$U(x,t) = \frac{A'(t)}{A(t)} + \int_0^x \frac{\partial}{\partial t}(tT(\xi,t))\mathrm{d}\xi \tag{1.58}$$

and also differentiate (1.57) with respect to x, which leads to

$$tT(x,t) = \int_0^t \frac{\partial}{\partial x}U(x,\tau)\mathrm{d}\tau. \tag{1.59}$$

We now substitute (1.59) into Eq. (1.58) and then put this new expression for $U(x,t)$ into Eq. (1.56), which leads to

$$H(x,t) = A(t)\exp\left[\int_0^t \int_0^x \frac{\partial}{\partial \xi}U(\xi,\tau)\xi\mathrm{d}\tau\right]. \tag{1.60}$$

We see that the exponent in Eq. (1.60) is a polynomial in x of degree $k+1$ if and only if the polynomial of maximal degree in $\{U_n(x)\}_{n=1}^\infty$ is of degree $k+1$, i.e., if and only if the set $\{P_n(x)\}_{n=0}^\infty$ for which the sequence $\{U_n(x)\}_{n=1}^\infty$ corresponds to is of C-Type k. Under this assumption, the exponent in Eq. (1.60) has the following structure:

$$\tilde{U}_1(x)t + \tilde{U}_2(x)t^2 + \cdots + \tilde{U}_{k+1}(x)t^{k+1},$$

which is the same form as the exponent in Eq. (1.48) when it corresponds to a B-Type 0 set. Thus, Eq. (1.60) can therefore be reduced to Eq. (1.54) using the same type of manipulation that was used in the previous section. Hence, we have proven the following theorem.

Theorem 1.16. *A set $\{P_n(x)\}_{n=0}^\infty$ is of C-Type k if and only if it is of B-Type k.*

1.4.4 A Summary of the Rainville σ-Type Classification

To complete our summary of *Type*, we conclude with a natural extension of Sheffer's classification, entitled σ-*Type*, which was constructed by E.D. Rainville and originally appeared in [8]. Now, we have seen that various results were ascertained from the generating function (1.19), which we now know characterizes A-*Type 0* sets. Moreover, we know that for $y = \exp[xH(t)]$ with $D := \mathrm{d}/\mathrm{d}x$ we have

$$Dy = H(t)y.$$

We intend to then modify the operator D so that it behaves in a similar fashion and also obtain a generating relation for the most basic type of sets classified by this modified operator. That is, if we replace D above by another differential operator, say σ, and the exponential function $\exp(z)$ by another function, e.g., $F(z)$, then we want to find σ such that

$$\sigma F(z) = F(z).$$

From there, we can construct an analogue of Eq. (1.19) in Theorem 1.5. This leads to the following definition of our differential operator σ (using the same notation as in [8]).

Definition 1.12. We define the differential operator σ as follows:

$$\sigma := D \prod_{i=1}^{q} (\theta + b_i - 1), \quad D := \frac{d}{dx} \text{ and } \theta := xD,$$

with $b_i \neq 0$ and b_i not equal to a negative integer.

In fact, this operator is a degree-lowering operator. To facilitate its application on polynomials and its degree-lowering nature, consider acting on a monomial, like x^2, with $q = 1$ and $b_1 = 1$. The function of the q and b_i-terms will become more transparent in Theorem 1.17 at the end of this section.

We now define σ-*Type*.

Definition 1.13. Let $\{P_n(x)\}_{n=0}^{\infty}$ be a set such that

$$J_\sigma [P_n(x)] = \sum_{k=0}^{\infty} T_k(x) \sigma^{k+1} P_n(x) = P_{n-1}(x),$$

with $\deg(T_j(x)) \leq k$. If the polynomial of maximal degree in the set of coefficients $\{T_j(x)\}_{j=0}^{\infty}$ is of degree k, then we classify $\{P_n(x)\}_{n=0}^{\infty}$ as σ-*Type k*. If the coefficients $\{T_j(x)\}_{j=0}^{\infty}$ are unbounded, then $\{P_n(x)\}_{n=0}^{\infty}$ is σ-*Type* ∞.

Based on this definition, we see that polynomials of σ-*Type 0* satisfy the following form:

$$J_\sigma [P_n(x)] = \sum_{k=0}^{\infty} c_k \sigma^{k+1} P_n(x) = P_{n-1}(x),$$

where each c_k is a constant and $c_0 \neq 0$. Since each of the c_k's is a constant, analogous to our A-*Type 0* analysis, there exists a generating function for J_σ, which we call $J_\sigma(t)$, and a corresponding inverse $H_\sigma(t)$ such that $J_\sigma(H_\sigma(t)) = H_\sigma(J_\sigma(t)) = t$:

$$J_\sigma(t) = \sum_{n=0}^{\infty} c_n t^{n+1}, \quad c_0 \neq 0 \quad \text{and} \quad H_\sigma(t) = \sum_{n=0}^{\infty} h_n t^{n+1}, \quad h_0 \neq 0.$$

This leads us to the main characterization theorem for σ-*Type 0* sets.

Theorem 1.17. *A set is of σ-Type 0, with σ as defined in Definition 1.12, if and only if* $\{P_n(x)\}_{n=0}^{\infty}$ *satisfies the generating function*

$$A(t){}_0F_q(-;b_1,b_2,\cdots,b_q;xH(t)) = \sum_{n=0}^{\infty} P_n(x)t^n$$

with $H(t)$ as defined above and $A(t)$ as in Eq. (1.20).

In Theorem 1.17, we wrote the generating function in the **generalized hypergeometric form**, which is defined as

$$_rF_s\left(\begin{array}{c}a_1,\ldots,a_r\\b_1,\ldots,b_s\end{array}\middle|\;x\right) = \sum_{k=0}^{\infty} \frac{(a_1,\ldots,a_r)_k}{(b_1,\ldots,b_s)_k} \frac{x^k}{k!}, \tag{1.61}$$

where the **Pochhammer symbol** $(a)_k$ is

$$(a)_k := a(a+1)(a+2)\cdots(a+k-1), \quad (a)_0 := 1, \tag{1.62}$$

and

$$(a_1,\ldots,a_j)_k := (a_1)_k \ldots (a_j)_k.$$

We now see that the selection of q is dependent on the number of denominator parameters (the b_i's) in the generating function of Theorem 1.17. For a proof of Theorem 1.17, the interested reader can refer to [8].

1.5   A Brief Discussion of Meixner's Analysis

In 1934, J. Meixner published [7] (written in German) in which he considered the generating relation (1.19) (with the same assumptions on the $A(t)$ and $H(t)$ as in Sheffer's work [9]) to be the *definition* of a certain class of polynomials. From there, he determined all sets that satisfy this relation that were also orthogonal and reached the same conclusions as Sheffer did in [9]. In other words, Meixner determined all orthogonal sets $\{P_n(x)\}_{n=0}^{\infty}$ that satisfy

$$f(t)e^{xu(t)} = \sum_{n=0}^{\infty} \frac{P_n(x)}{n!}t^n; \quad f(0) = 1, \; u(0) = 0, \; \frac{\mathrm{d}}{\mathrm{d}t}u(0) = 1, \tag{1.63}$$

which we have written using Meixner's notation, which we essentially adhere to throughout this section.

In a similar manner as Sheffer, Meixner defined a general degree-lowering, linear, differential operator of infinite order, which we call J as opposed to his "t" to avoid confusion with t-parameters, which satisfies a certain commutation relation with the differential operator D. Moreover, $J(t)$ is a formal power series without a constant term and with a unitary linear coefficient. The formal power series inverse of $J(t)$ was defined to be $u(t)$, i.e., $u(J(t)) = J(u(t)) = t$.

Throughout his work, Meixner considered $\{P_n(x)\}_{n=0}^{\infty}$ to be a set of monic polynomials. With this convention, he utilized the following monic three-term recurrence relation:

$$P_{n+1}(x) = (x + l_{n+1})P_n(x) + k_{n+1}P_{n-1}(x), \quad n = 0, 1, 2, \ldots \quad (1.64)$$

with $l_{n+1} \in \mathbb{R}$ and $k_{n+1} \in \mathbb{R}^-$ and demonstrated that

$$J(D)P_n(x) = nP_{n-1}(x). \quad (1.65)$$

Now, we act on Eq. (1.64) with $J(D)$ as in Eq. (1.65) and obtain

$$(n+1)P_n(x) = (x + l_{n+1})nP_{n-1}(x) + J'(D)P_n(x) + k_{n+1}(n-1)P_{n-2}(x), \quad (1.66)$$

where $J'(D)$ is of course the derivative of the formal power series $J(D)$. Then, we replace n with $n-1$ in Eq. (1.64) and multiply both sides by n to obtain

$$nP_n(x) = (x + l_n)nP_{n-1}(x) + k_n nP_{n-2}(x). \quad (1.67)$$

Next, we subtract (1.67) from Eq. (1.66) which yields

$$\left(1 - J'(D)\right)P_n(x) = (l_{n+1} - l_n)nP_{n-1}(x) + \left(\frac{k_{n+1}}{n} - \frac{k_n}{n-1}\right)n(n-1)P_{n-2}(x).$$

We then replace n with $n+1$ in the recursion coefficients above leading to

$$\left(1 - J'(D)\right)P_n(x) = (l_{n+2} - l_{n+1})nP_{n-1}(x) + \left(\frac{k_{n+2}}{n+1} - \frac{k_{n+1}}{n}\right)n(n-1)P_{n-2}(x).$$

We assign

$$\lambda := l_{n+1} - l_n, \quad n = 1, 2, \ldots, \quad (1.68)$$

$$\kappa := \frac{k_{n+1}}{n} - \frac{k_n}{n-1}, \quad n = 2, 3, \ldots \quad (1.69)$$

giving

$$\left(1 - J'(D)\right)P_n(x) = \lambda nP_{n-1}(x) + \kappa n(n-1)P_{n-2}(x), \quad n = 0, 1, 2, \ldots$$

and via Eq. (1.65) we see that this recurrence can be written as

$$\left(1 - J'(D)\right)P_n(x) = \lambda J(D)P_n(x) + \kappa J^2(D)P_n(x), \quad n = 0, 1, 2, \ldots \quad (1.70)$$

From rewriting Eqs. (1.68) and (1.69) as

$$l_{n+1} = l_n + \lambda \quad \text{and} \quad k_{n+1} = n\left(\frac{k_n}{n-1} + \kappa\right)$$

and iterating, we see that $l_{n+1} = l_1 + n\lambda$ and $k_{n+1} = k_2 + (n-1)\kappa$. We then substitute these recursion coefficients into Eq. (1.64) to obtain the following three-term recurrence relation:

$$P_{n+1}(x) = (x + l_1 + n\lambda)P_n(x) + n(k_2 + (n-1)\kappa)P_{n-1}(x), \quad n = 0, 1, 2, \ldots \quad (1.71)$$

with $k_2 < 0$ and $\kappa \le 0$ from the original restrictions imposed upon Eq. (1.64).

Now, from Eq. (1.70), it follows that

$$J'(u(t)) = 1 - \lambda t - \kappa t^2.$$

Differentiating both sides of the relation $J(u(t)) = t$ tells us that $J'(u(t)) = 1/u'(t)$. Thus, our relation directly above becomes

$$J'(u(t)) = 1 - \lambda t - \kappa t^2 = \frac{1}{u'(t)}. \quad (1.72)$$

By setting $x = 0$, we note that the generating function (1.63) turns into

$$f(t) = \sum_{n=0}^{\infty} \frac{P_n(0)t^n}{n!}.$$

Thus, multiplying both sides of Eq. (1.71) by $t^n/n!$, setting $x = 0$ and summing for $n = 0, 1, 2, \ldots$ lead to the differential equation

$$\frac{f'(t)}{f(t)} = \frac{k_2 t}{1 - \lambda t - \kappa t^2}.$$

Factoring $1 - \lambda t - \kappa t^2$ as $(1 - \alpha t)(1 - \beta t)$ gives

$$\frac{f'(t)}{f(t)} = \frac{k_2 t}{(1 - \alpha t)(1 - \beta t)}, \quad \alpha, \beta \in \mathbb{C}. \quad (1.73)$$

We can now exhaust every possible combination of α and β (and incorporate λ and κ as well), substitute each of them into Eqs. (1.72) and (1.73), and solve the resulting differential equations. In each case, the solution to Eq. (1.72) will yield an expression for $u(t)$ and the solution to Eq. (1.73) will yield an expression for $f(t)$. In substituting these into Eq. (1.63), we achieve a generating function for each orthogonal set.

Below, we write the results for each of these aforementioned cases. In each case, part (i) denotes the solutions to Eqs. (1.72) and (1.73), which respectively yield our expressions for $u(t)$ and $f(t)$. In part (ii), we write each of the resulting generating functions in their rescaled form, so they appear as they do in the contemporary literature.

Case I. The Hermite Polynomials: $\alpha = \beta = 0$ $(\lambda = \kappa = 0)$

$$(i) \ \ u(t) = t \quad \text{and} \quad f(t) = \exp\left(k_2 t^2 / 2\right)$$

$$(ii) \ \ \exp\left(2xt - t^2\right) = \sum_{n=0}^{\infty} \frac{H(x)}{n!} t^n$$

Case II. The Laguerre Polynomials: $\alpha = \beta \neq 0$

$$(i) \ \ u(t) = \frac{t}{1 - \alpha t} \quad \text{and} \quad f(t) = (1 - \alpha t)^{k_2/\alpha^2} \exp\left(\frac{t}{1 - \alpha t} \frac{k_2}{\alpha}\right)$$

$$(ii) \ \ (1 - t)^{-(\alpha+1)} \exp\left(\frac{xt}{t - 1}\right) = \sum_{n=0}^{\infty} L_n^{(\alpha)}(x) t^n$$

Case III. The Charlier Polynomials: $\alpha \neq 0$ and $\beta = 0$ $(\kappa = 0)$

$$(i) \ \ u(t) = -\frac{1}{\alpha} \ln(1 - \alpha t) \quad \text{and} \quad f(t) = (1 - \alpha t)^{-k_2/\alpha^2} e^{-k_2 t/\alpha}$$

$$(ii) \ \ e^t \left(1 - \frac{t}{\alpha}\right)^x = \sum_{n=0}^{\infty} \frac{C_n(x; \alpha)}{n!} t^n$$

Case IV. Meixner determined *two* orthogonal sets that stem from this case. The general $u(t)$ and $f(t)$ are as follows:

$$u(t) = \frac{1}{\alpha - \beta} \ln\left(\frac{1 - \beta t}{1 - \alpha t}\right) \quad \text{and} \quad f(t) = \left(\frac{(1 - \beta t)^{1/\beta}}{(1 - \alpha t)^{1/\alpha}}\right)^{k_2/(\alpha - \beta)}.$$

(a) **The Meixner Polynomials**: $\alpha \neq \beta$ and $\alpha, \beta \in \mathbb{R}$ $(\kappa \neq 0)$

$$\left(1 - \frac{t}{c}\right)^x (1 - t)^{-(x+\beta)} = \sum_{n=0}^{\infty} \frac{(\beta)_n}{n!} M(x; \beta, c) t^n.$$

(b) **The Meixner–Pollaczek Polynomials**: $\alpha \neq \beta$, α and β complex conjugates $(\kappa \neq 0)$

$$(1 - e^{i\phi}t)^{-\lambda+ix}(1 - e^{-i\phi}t)^{-\lambda-ix} = \sum_{n=0}^{\infty} P_n^{(\lambda)}(x; \phi)t^n.$$

Remark 1.1. The Krawtchouk polynomials are the *third* orthogonal set that comes from Case IV above. These polynomials were not included in either Meixner's or Sheffer's work.

Example 1.5. As a simple example of the scaling process, we see that in Case I above, $u(t) = t$ and $f(t) = \exp(k_2 t^2/2)$, so that

$$f(t)e^{xu(t)} = \exp\left(xt + \frac{1}{2}k_2 t^2\right).$$

Thus, we can obtain the generating relation in Case I by simply choosing $k_2 = -1/2$ and rescaling t via $t \to 2t$.

1.5.1 Al-Salam's Extension of Meixner's Characterization

To briefly supplement our discussion of Meixner's analysis, we state that W. A. Al-Salam extended the results of Meixner, and therefore Sheffer, in [1]. Namely, he showed that the left-hand side of Eq. (1.63) can be replaced with $\exp(Q(x,t))$, where $Q(x,t)$ is a polynomial in x with coefficients that are functions of t, as seen below:

$$\exp(Q(x,t)) = \sum_{n=0}^{\infty} P_n(x)\frac{t^n}{n!},$$

$$Q(x,t) = \sum_{j=0}^{k} x^j a^{(j)}(t), \quad k \geq 1, \quad a^{(j)}(t) = \sum_{r=0}^{\infty} a_r^{(j)}t^r, \quad j = 0, 1, 2, \ldots, k$$

and that the resulting orthogonal polynomials $\{P_n(x)\}_{n=0}^{\infty}$ will be the *same* as those achieved by Meixner and Sheffer. This showed that the conditions on the generating function (1.63) can be weakened without yielding new orthogonal sets.

References

1. W.A. Al-Salam, *On a Characterization of Meixner's Polynomials*, Quart. J. Math. Oxford, 17(1966), 7–10.
2. W.A. Al-Salam, *Characterization theorems for orthogonal polynomials*, in: Nevai, P. (Ed.), Orthogonal Polynomials: Theory and Practice. Kluwer Academic Publishers, Dordrecht, pp. 1–24, (1990).

3. G.E. Andrews, R.A. Askey and R. Roy, *Special Functions*, Encyclopedia of Mathematics and its Applications 71, Cambridge University Press, Cambridge, 1999.
4. G. Gasper and M. Rahman, *Basic Hypergeometric Series*, second ed., Encyclopedia of Mathematics and its Applications 96, Cambridge University Press, Cambridge, 2004.
5. M.E.H. Ismail, *Classical and Quantum Orthogonal Polynomials in One Variable*, Encyclopedia of Mathematics and its Applications 98, Cambridge University Press, Cambridge, 2005.
6. R. Koekoek and R.F. Swarttouw, *The Askey-scheme of hypergeometric orthogonal polynomials and its q-analogue*, Reports of the Faculty of Technical Mathematics and Information, No. 98–17, Delft University of Technology, (1998). http://aw.twi.tudelft.nl/~koekoek/askey/index.html
7. J. Meixner, *Orthogonale Polynomsysteme mit einer besonderen Gestalt der erzeugenden Funktion*, J. London Math. Soc., 9(1934), 6–13.
8. E.D. Rainville, *Special Functions*, Macmillan, New York, 1960.
9. I.M. Sheffer, *Some properties of polynomial sets of type zero*, Duke Math J., 5(1939), 590–622.
10. G. Szegő, *Orthogonal Polynomials*, fourth ed., American Mathematical Society, Colloquium Publications, Vol. XXIII, Providence, 1975.

Chapter 2
Some Applications of the Sheffer A-Type 0 Orthogonal Polynomial Sequences

In this chapter, we address several of the many applications of the classical orthogonal polynomial sequences. These applications include first-order differential equations that characterize linear generating functions, additional first-order differential equations, second-order differential equations (with applications to quantum mechanics), difference equations and numerical integration (Gaussian Quadrature). We first develop each of these applications in a general context and then cover examples using specific Sheffer Sequences, i.e. the Laguerre, Hermite, Charlier, Meixner, Meixner–Pollaczek, and Krawtchouk polynomials.

2.1 Preliminaries

Throughout this chapter, we make use of each of the following definitions, terminologies and notations.

Definition 2.1. We always assume that a *set* of polynomials $\{P_n(x)\}_{n=0}^{\infty}$ is such that each $P_n(x)$ has degree exactly n, which we write as $\deg(P_n(x)) = n$.

Definition 2.2. A set of polynomials $\{Q_n(x)\}_{n=0}^{\infty}$ is *monic* if $Q_n(x) - x^n$ is of degree at most $n-1$ or equivalently if the leading coefficient of each $Q_n(x)$ is unitary.

Definition 2.3. The set of polynomials $\{P_n(x)\}_{n=0}^{\infty}$ is *orthogonal* if it satisfies one of the two weighted inner products below:

$$\text{Continuous}: \quad \langle P_m(x), P_n(x) \rangle = \int_{\Omega_1} P_m(x)P_n(x)w(x)\mathrm{d}x = \alpha_n \delta_{m,n}, \quad (2.1)$$

$$\text{Discrete}: \quad \langle P_m(x), P_n(x) \rangle = \sum_{\Omega_2} P_m(x)P_n(x)w(x) = \beta_n \delta_{m,n}, \quad (2.2)$$

where $\delta_{m,n}$ denotes the *Kronecker delta*

D.J. Galiffa, *On the Higher-Order Sheffer Orthogonal Polynomial Sequences*,
SpringerBriefs in Mathematics, DOI 10.1007/978-1-4614-5969-9_2,
© Daniel J. Galiffa 2013

$$\delta_{m,n} := \begin{cases} 1 & \text{if} \quad m = n \\ 0 & \text{if} \quad m \neq n, \end{cases}$$

with $\Omega_1 \subseteq \mathbb{R}$, $\Omega_2 \subseteq \{0, 1, 2, \ldots\}$, and $w(x) > 0$ is entitled the **weight function**. We also always assume the following **normalizations**:

$$\int_{\Omega_1} w(x)\mathrm{d}x = 1 \quad \text{and} \quad \sum_{\Omega_2} w(x) = 1.$$

Definition 2.4 (The Three-Term Recurrence Relations). It is a necessary and sufficient condition that an orthogonal polynomial sequence $\{P_n(x)\}_{n=0}^{\infty}$ satisfies an **unrestricted three-term recurrence relation** of the form

$$P_{n+1}(x) = (A_n x + B_n)P_n(x) - C_n P_{n-1}(x), \quad A_n A_{n-1} C_n > 0,$$

where $P_{-1}(x) = 0$ and $P_0(x) = 1$. (2.3)

If $Q_n(x)$ represents the monic form of $P_n(x)$, then it is a necessary and sufficient condition that $\{Q_n(x)\}_{n=0}^{\infty}$ satisfies the following **monic three-term recurrence relation**

$$Q_{n+1}(x) = (x - b_n)Q_n(x) - c_n Q_{n-1}(x), \quad c_n > 0,$$

where $Q_{-1}(x) = 0$ and $Q_0(x) = 1$. (2.4)

We entitle the conditions $A_n A_{n-1} C_n > 0$ and $c_n > 0$ above **positivity conditions**.

Definition 2.5. We shall define a **generating function** for a polynomial sequence $\{P_n(x)\}_{n=0}^{\infty}$ as follows:

$$\sum_{\Lambda} \zeta_n P_n(x)t^n = F(x,t),$$

with $\Lambda \subseteq \{0, 1, 2, \ldots\}$ and $\{\zeta_n\}_{n=0}^{\infty}$ a sequence in n that is independent of x and t. Moreover, we say that the function $F(x,t)$ **generates** the set $\{P_n(x)\}_{n=0}^{\infty}$.

Before we give our next definition, we discuss that in 1939 Sheffer [22] developed a characterization theorem that gave necessary and sufficient conditions for a polynomial sequence to be *A-Type 0* via a linear generating function. Originally, in 1934, J. Meixner published [15], wherein he essentially determined which orthogonal sets satisfy the aforementioned *A-Type 0* generating function using a different approach than Sheffer. Meixner basically used the *A-Type 0* generating function as the *definition* of the *A-Type 0* class. In this chapter, we follow Meixner's convention. We mention that the interested reader can also refer to [1] for a concise overview of Meixner's analysis.

Definition 2.6. A polynomial set $\{P_n(x)\}_{n=0}^{\infty}$ is classified as *A-Type 0* if $\{a_j\}_{j=0}^{\infty}$ and $\{h_j\}_{j=1}^{\infty}$ exist such that

$$A(t)e^{xH(t)} = \sum_{n=0}^{\infty} P_n(x)t^n,$$

$$A(t) := \sum_{n=0}^{\infty} a_n t^n, \quad a_0 = 1 \quad \text{and} \quad H(t) := \sum_{n=1}^{\infty} h_n t^n, \quad h_1 = 1.$$

The orthogonal sets that satisfy Definition 2.6, which are often simply called the **Sheffer Sequences**, are listed below as defined by their *A-Type 0* generating function.

The Laguerre Polynomials $\{L_n^{(\alpha)}(x)\}_{n=0}^{\infty}$

$$\sum_{n=0}^{\infty} L_n^{(\alpha)}(x)t^n = (1-t)^{-(\alpha+1)} \exp\left(\frac{xt}{t-1}\right).$$

The Hermite Polynomials $\{H_n(x)\}_{n=0}^{\infty}$

$$\sum_{n=0}^{\infty} \frac{1}{n!} H_n(x)t^n = \exp(2xt - t^2).$$

The Charlier Polynomials $\{C_n(x;a)\}_{n=0}^{\infty}$

$$\sum_{n=0}^{\infty} \frac{1}{n!} C_n(x;a)t^n = e^t \left(1 - \frac{t}{a}\right)^x.$$

The Meixner Polynomials $\{M(x;\beta,c)\}_{n=0}^{\infty}$

$$\sum_{n=0}^{\infty} \frac{(\beta)_n}{n!} M(x;\beta,c)t^n = \left(1 - \frac{t}{c}\right)^x (1-t)^{-(x+\beta)}.$$

The Meixner–Pollaczek Polynomials $\{P_n^{(\lambda)}(x;\phi)\}_{n=0}^{\infty}$

$$\sum_{n=0}^{\infty} P_n^{(\lambda)}(x;\phi)t^n = (1 - e^{i\phi}t)^{-\lambda+ix}(1 - e^{-i\phi}t)^{-\lambda-ix}.$$

The Krawtchouk Polynomials $\{K_n(x;p,N)\}_{n=0}^{\infty}$

$$\sum_{n=0}^{N} C(N,n)K_n(x;p,N)t^n = \left(1 - \frac{1-p}{p}t\right)^x (1+t)^{N-x},$$

for $x = 0,1,2,\dots,N$, where $C(N,n)$ denotes the binomial coefficient.

Example 2.1. We see that we can write the generating function for the Krawtchouk polynomials as

$$\sum_{n=0}^{\infty} \frac{1}{n!} K_n(x; p, N) t^n = (1+t)^N \exp\left(x \ln\left(\frac{(p-1)t+p}{p(1+t)}\right)\right)$$

from which $A(t)$ and $H(t)$ can be readily identified. Similar trivial manipulations can be made to the generating functions of the remaining five orthogonal sets to obtain the form $A(t) \exp(xH(t))$.

Now that both orthogonality and the Sheffer Sequences have been defined, we address the fact that the Laguerre, Hermite, and Meixner–Pollaczek polynomials satisfy a continuous orthogonality relation of the form (2.1), and the Charlier, Meixner, and Krawtchouk polynomials satisfy a discrete orthogonality relation of the form (2.2). For more information refer to [11] and the references therein.

Definition 2.7. We can express each of our polynomials in the *generalized hypergeometric form* $(_rF_s)$ as seen below:

$$_rF_s\left(\begin{matrix} a_1, \dots, a_r \\ b_1, \dots, b_s \end{matrix} \middle| z\right) = \sum_{k=0}^{\infty} \frac{(a_1, \dots, a_r)_k}{(b_1, \dots, b_s)_k} \frac{z^k}{k!}, \tag{2.5}$$

where the *Pochhammer symbol* $(a)_k$ is defined as

$$(a)_k := a(a+1)(a+2)\cdots(a+k-1), \quad (a)_0 := 1 \tag{2.6}$$

and

$$(a_1, \dots, a_j)_k := (a_1)_k \cdots (a_j)_k.$$

The sum (2.5) terminates if one of the numerator parameters is a negative integer, e.g., if one such parameter is $-n$, then (2.5) is a finite sum on $0 \leq k \leq n$.

2.2   Differential Equations Part I: The "Inverse Method"

In this section, we demonstrate how each of the *A-Type 0* generating functions satisfy a first-order differential equation. We entitle our approach for deriving these differential equations *"the inverse method"* because of the connection our approach has to inverse problems. That is, in the study of orthogonal polynomials, the term *inverse problem* refers to the problem of obtaining the weight function of an orthogonal set by using only the corresponding recursion coefficients. For inverse problems, the generating function that is obtained via a differential equation can be viewed as a by-product. For additional examples of inverse problems, consider Chap. 5 of [12] and the references therein.

To begin, we assume that $\{P_n(x)\}_{n=0}^{\infty}$ is a polynomial set that satisfies an unrestricted three-term recurrence relation of the form (2.3). We first multiply this relation by $c_n t^n$, where c_n is a function in n that is independent of x and t, and sum for $n = 0, 1, 2, \ldots$ Then, from the assignment $F(t;x) := \sum_{n=0}^{\infty} c_n P_n(x) t^n$, we obtain a first-order differential equation in t with x regarded as a parameter. The initial condition for this equation is $F(0;x) = 1$ via the initial condition $P_0(x) = 1$ in (2.3). The existence and uniqueness of the solution to this differential equation is ensured and the solution $F(t;x)$ will be a generating function for the set $\{P_n(x)\}_{n=0}^{\infty}$. To demonstrate the procedure, we work out all of the details for the Charlier and Laguerre polynomials and sketch the details for the Miexner–Pollaczek polynomials.

Example 2.2. To begin, we note that from examining the generating function of the Charlier polynomials, c_n as described above must be $1/n!$. The three-term recurrence relation for the Charlier polynomials can be written as

$$-xC_n(x;a) = aC_{n+1}(x;a) - (n+a)C_n(x;a) + nC_{n-1}(x;a).$$

Thus, we multiply both sides of this relation by $t^n/n!$ and sum the result for $n = 0, 1, 2, \ldots$:

$$-x\sum_{n=0}^{\infty} \frac{C_n(x;a)}{n!} t^n = a\sum_{n=0}^{\infty} \frac{C_{n+1}(x;a)}{n!} t^n - \sum_{n=1}^{\infty} \frac{C_n(x;a)}{(n-1)!} t^n$$

$$- a\sum_{n=0}^{\infty} \frac{C_n(x;a)}{n!} t^n + \sum_{n=1}^{\infty} \frac{C_{n-1}(x;a)}{(n-1)!} t^n.$$

Next, we define $F := F(t;x,a) := \sum_{n=0}^{\infty} \frac{C_n(x;a)}{n!} t^n$ and it therefore follows that

$$\dot{F} := \frac{\partial}{\partial t} F(t;x,a) = \sum_{n=1}^{\infty} \frac{C_n(x;a)}{(n-1)!} t^{n-1}.$$

Then, we see that our relation becomes

$$\dot{F} - \left(1 + \frac{x}{t-a}\right) F = 0,$$

which is a first-order differential equation with initial condition $F(0;x,a) = 1$. A general solution is

$$F(t;x,a) = c(x;a) e^t (a-t)^x,$$

where $c(x;a)$ is an arbitrary function of x. From the initial condition, it is immediate that $c(x;a) = a^{-x}$ and thus, the unique solution turns out to be

$$F(t;x,a) = \sum_{n=0}^{\infty} \frac{C_n(x;a)}{n!} t^n = e^t \left(1 - \frac{t}{a}\right)^x,$$

which is the Sheffer *A-Type 0* generating function for the Charlier polynomials.

Example 2.3. We next consider the Laguerre polynomials, which have the following unrestricted three-term recurrence relation:

$$(n+1)L_{n+1}^{(\alpha)}(x) - (2n+\alpha+1-x)L_n^{(\alpha)}(x) + (n+\alpha)L_{n-1}^{(\alpha)}(x) = 0.$$

We multiply both sides of this relation by t^n ($c_n \equiv 1$) and sum for $n = 0,1,2,\ldots$, which yields

$$\sum_{n=0}^{\infty}(n+1)L_{n+1}^{(\alpha)}(x)t^n - 2\sum_{n=1}^{\infty}nL_n^{(\alpha)}(x)t^n - (\alpha+1-x)\sum_{n=0}^{\infty}L_n^{(\alpha)}(x)t^n$$

$$+ \sum_{n=1}^{\infty}nL_{n-1}^{(\alpha)}(x)t^n + \alpha\sum_{n=0}^{\infty}L_{n-1}^{(\alpha)}(x)t^n = 0.$$

We next assign $G := G(t;x) := \sum_{n=0}^{\infty}L_n^{(\alpha)}(x)t^n$ and recall that $L_{-1}^{(\alpha)}(x) = 0$, which gives

$$\dot{G} - 2t\dot{G} - (\alpha+1-x)G + \sum_{n=2}^{\infty}nL_{n-1}^{(\alpha)}(x)t^n + \alpha tG = 0.$$

We also observe that

$$\sum_{n=1}^{\infty}nL_{n-1}^{(\alpha)}(x)t^n = \sum_{n=2}^{\infty}(n-1)L_{n-1}^{(\alpha)}(x)t^n + \sum_{n=1}^{\infty}L_{n-1}^{(\alpha)}(x)t^n = t^2\dot{G}+tG.$$

Using all of this, we can put our relation into standard form:

$$\dot{G} + \left[\frac{x+(\alpha+1)(t-1)}{1-2t+t^2}\right]G = 0; \quad G(0;x) = 1.$$

The integrating factor in this equation turns out to be $\mu = \exp\left[\int \frac{x+(\alpha+1)(t-1)}{1-2t+t^2}dt\right]$ and through partial fraction decomposition, we obtain the general solution

$$G(t;x) = c(x,\alpha)(t-1)^{-(\alpha+1)}\exp\left(\frac{x}{t-1}\right).$$

Therefore, from using our initial condition to determine $c(x,\alpha)$, we establish the solution

$$G(t;x) = \sum_{n=0}^{\infty}L_n^{(\alpha)}(x)t^n = (t-1)^{-(\alpha+1)}\exp\left(\frac{xt}{t-1}\right),$$

which is the Sheffer *A-Type 0* generating function for the Laguerre polynomials.

Example 2.4. For a more detailed example, we now consider the Meixner–Pollaczek polynomials. These polynomials have the unrestricted three-term recurrence relation

$$(n+1)P_{n+1}^{(\lambda)}(x;\phi) - 2\left[x\sin\phi + (n+\lambda)\cos\phi\right]P_n^{(\lambda)}(x;\phi) + (n+2\lambda-1)P_{n-1}^{(\lambda)}(x;\phi) = 0.$$

We then multiply this relation by t^n, take $c_n \equiv 1$ and sum for $n = 0, 1, 2, \ldots$. From letting $H := H(t;x,\lambda,\phi) := \sum_{n=0}^{\infty} P_n^{(\lambda)}(x;\phi)t^n$ we achieve the differential equation

$$\dot{H} + 2\left(\frac{\lambda(t-\cos\phi) - x\sin\phi}{1 - 2\cos\phi\, t + t^2}\right)H = 0; \quad H(0;x,\lambda,\phi) = 1.$$

The solution to this equation can be obtained by partial fraction decomposition, and several manipulations, to be

$$H(t;x,\lambda,\phi) = \sum_{n=0}^{\infty} P_n^{(\lambda)}(x;\phi)t^n = (1 - e^{i\phi}t)^{-\lambda+ix}(1 - e^{-i\phi}t)^{-\lambda-ix}.$$

2.3   Differential Equations Part II

We now discuss some additional characterizations of classical orthogonal sets via differential equations. These results lead to a way to solve the time-independent Schrödinger equation. Through our development of each characterization, additional results, definitions and concepts are addressed, which are important unto themselves. We also supplement our characterizations by covering specific details for the Laguerre polynomials and discuss how similar results can be achieved for other *A-Type 0* sets. This section is based on much of the analysis conducted in [3–5, 7, 9, 16, 21, 23]. We begin with an important fundamental condition.

Lemma 2.1. *An Equivalent Orthogonality Condition: The set* $\{P_n(x)\}_{n=0}^{\infty}$ *is orthogonal with respect to the weight function* $w(x) > 0$ *on* Ω_1 *if and only if*

$$\int_{\Omega_1} x^j P_n(x)w(x)dx = 0, \quad \forall j = 0, 1, 2, \ldots, n-1. \tag{2.7}$$

Proof. ($\Rightarrow$) Assume (2.7) holds. It is clear that there exist constants $\{\eta_{m,j}\}$ such that

$$P_m(x) = \sum_{j=0}^{m} \eta_{m,j}x^j.$$

We first assume that $m < n$, in which case we have

$$\int_{\Omega_1} P_m(x)P_n(x)w(x)dx = \sum_{j=0}^{m} \eta_{m,j}\int_{\Omega_1} x^j P_n(x)w(x)dx = 0,$$

because $j \leq m < n$. If $m > n$, we can simply interchange m with n in the logic used above. Hence, we have shown that given (2.7) it necessarily follows that

$$\int_{\Omega_1} P_m(x)P_n(x)w(x)dx = 0, \text{ for } m \neq n,$$

and therefore $\{P_n(x)\}_{n=0}^{\infty}$ is an orthogonal set with respect to the weight function $w(x) > 0$ on Ω_i.

($\Leftarrow$) Assume the orthogonality relation directly above is true. Then, there exist constants $\{\tilde{\eta}_{m,j}\}$ such that

$$x^j = \sum_{m=0}^{j} \tilde{\eta}_{m,j}P_m(x).$$

Therefore, for every $j \in \{0, 1, \ldots, n-1\}$, we know that

$$\int_{\Omega_1} x^j P_n(x)w(x)dx = \sum_{m=0}^{j} \tilde{\eta}_{m,j} \int_{\Omega_1} P_m(x)P_n(x)w(x)dx = 0,$$

because $m \leq j < n$, i.e., since $m \neq n$. $\square$

We will also need the following result.

Lemma 2.2. *With α_n as in (2.1) and c_n as in (2.4), we have*

$$\alpha_n = \prod_{k=1}^{n} c_k.$$

Proof. We first multiply both sides of (2.4) by $P_{n-1}(x)w(x)$ and integrate over Ω_1, leading to

$$\int_{\Omega_1} P_{n+1}(x)P_{n-1}(x)w(x)dx = \int_{\Omega_1} xP_n(x)P_{n-1}(x)w(x)dx$$

$$- b_n \int_{\Omega_1} P_n(x)P_{n-1}(x)w(x)dx$$

$$- c_n \int_{\Omega_1} P_{n-1}(x)P_{n-1}(x)w(x)dx.$$

From the orthogonality relation (2.1), we observe that our relation directly above becomes

$$c_n \alpha_{n-1} = \int_{\Omega_1} xP_n(x)P_{n-1}(x)w(x)dx.$$

Since our polynomial sequence $\{P_n(x)\}_{n=0}^{\infty}$ is monic, we have

$$c_n \alpha_{n-1} = \int_{\Omega_1} P_n(x) \left(P_n(x) + \mathcal{O}\left(x^{n-1}\right) \right) w(x) dx$$

$$= \int_{\Omega_1} P_n^2(x) w(x) dx$$

$$= \alpha_n,$$

which follows from Lemma 2.1 and (2.1). Thus, by iterating $c_n \alpha_{n-1} = \alpha_n$, with $\alpha_0 = 1$ via the normalization in Definition 2.3, we obtain our result. $\square$

Theorem 2.1 (The Christoffel–Darboux Identity). *For $N > 0$ we have*

$$\sum_{k=0}^{N-1} \frac{1}{\alpha_k} P_k(x) P_k(y) = \frac{P_N(x) P_{N-1}(y) - P_N(y) P_{N-1}(x)}{\alpha_{N-1}(x-y)}. \tag{2.8}$$

Proof. After replacing $Q_n(x)$ with $P_n(x)$ in (2.4), we multiply both sides of the resulting relation by $P_n(y)$, leading to

$$P_{n+1}(x) P_n(y) = x P_n(x) P_n(y) - b_n P_n(x) P_n(y) - c_n P_{n-1}(x) P_n(y). \tag{2.9}$$

We then exchange x with y in (2.9) and subtract this result from (2.9), which gives

$$P_{n+1}(x) P_n(y) - P_{n+1}(y) P_n(x)$$
$$= (x-y) P_n(x) P_n(y) + c_n \left(P_n(x) P_{n-1}(y) - P_n(y) P_{n-1}(x) \right). \tag{2.10}$$

We then define the following operator:

$$\Delta_k^{x,y} := P_k(x) P_{k-1}(y) - P_k(y) P_{k-1}(x)$$

and we see that (2.10) becomes

$$(x-y) P_n(x) P_n(y) = \Delta_{n+1}^{x,y} - c_n \Delta_n^{x,y}.$$

Lastly, we divide both sides of the equation above by α_n and utilize Lemma 2.2, which yields

$$\frac{1}{\alpha_n} P_n(x) P_n(y) = \frac{1}{x-y} \left[\frac{1}{\alpha_n} \Delta_{n+1}^{x,y} - \frac{1}{\alpha_{n-1}} \Delta_n^{x,y} \right].$$

Summing both sides of this identity from 0 to $N-1$, we see that the right-hand side becomes a telescoping series that converges to the right-hand side of (2.8). $\square$

The limiting case $y \to x$ of (2.8) is readily achieved via L'Hôpital's Rule to be

$$\sum_{k=0}^{N-1} \frac{1}{\alpha_k} P_k^2(x) = \frac{P_N'(x) P_{N-1}(x) - P_N(x) P_{N-1}'(x)}{\alpha_{N-1}}. \tag{2.11}$$

The Christoffel–Darboux Identity has many general usages and will be needed throughout this chapter.

We next write $w(x)$ as in (2.1) as

$$w(x) = \exp(-v(x)),\qquad(2.12)$$

where $v = v(x)$ is a twice continuously differentiable function on Ω_1. For this section, we also use the *orthonormal* form $\{p_n(x)\}_{n=0}^{\infty}$ of $\{P_n(x)\}_{n=0}^{\infty}$. This definition is as follows.

Definition 2.8 (Orthonormality). It is a necessary and sufficient condition that an orthogonal polynomial sequence $\{p_n(x)\}_{n=0}^{\infty}$ satisfies an **orthonormal three-term recurrence relation** of the form

$$x p_n(x) = a_{n+1}p_{n+1}(x) + b_n p_n(x) + a_n p_{n-1}(x),$$

$$\text{where } p_{-1}(x) = 0 \text{ and } p_0(x) = 1.\qquad(2.13)$$

For c_n as in (2.4) we have (see [9]):

$$a_n = \sqrt{c_n}.\qquad(2.14)$$

Therefore, we can of course equivalently write (2.13) as

$$x p_n(x) = \sqrt{c_{n+1}}\, p_{n+1}(x) + b_n p_n(x) + \sqrt{c_n}\, p_{n-1}(x).$$

The orthonormal form $p_n(x)$ can be expressed in terms of $P_n(x)$ by

$$p_n(x) = \frac{1}{\sqrt{\alpha_n}} P_n(x).\qquad(2.15)$$

This leads to the (continuous) **orthonormal relation**

$$\langle p_m(x), p_n(x)\rangle = \int_{\Omega_1} p_m(x)p_n(x)w(x)\mathrm{d}x = \delta_{m,n}.$$

Remark 2.1. We do not call upon a discrete orthonormal relation in this section.

We now have the following.

Theorem 2.2. *If $v(x)$ is as defined in (2.12) and $\{p_n(x)\}_{n=0}^{\infty}$ is an orthonormal set as defined by Definition 2.8, then $p_n(x)$ satisfies the differential equation*

$$p_n'(x) = -B_n(x)p_n(x) + A_n(x)p_{n-1}(x),\qquad(2.16)$$

with $A_n(x)$ and $B_n(x)$ as defined as

$$A_n(x) = a_n \left(\left. \frac{p_n^2(y)w(y)}{y-x} \right|_{\Omega_1} + \int_{\Omega_1} \frac{v'(x) - v'(y)}{x-y} p_n^2(y)w(y)\mathrm{d}y \right), \tag{2.17}$$

$$B_n(x) = a_n \left(\left. \frac{p_n(y)p_{n-1}(y)w(y)}{y-x} \right|_{\Omega_1} + \int_{\Omega_1} \frac{v'(x) - v'(y)}{x-y} p_n(y)p_{n-1}(y)w(y)\mathrm{d}y \right), \tag{2.18}$$

when the above integrals and boundary terms exist.

Proof. Since $\deg(p_n'(x)) = n-1$, there exist constants $\{c_{n,k}\}$ such that

$$p_n'(x) = \sum_{k=0}^{n-1} c_{n,k} p_k(x). \tag{2.19}$$

We then multiply both sides of (2.19) by $p_m(x)w(x)$ and integrate over Ω_1:

$$\int_{\Omega_1} p_n'(x)p_m(x)w(x)\mathrm{d}x = \sum_{k=0}^{n-1} c_{n,k} \int_{\Omega_1} p_k(x)p_m(x)w(x)\mathrm{d}x.$$

We then observe that the right-hand side is nonzero if and only if $m=k$, in which case we have

$$c_{n,k} = \int_{\Omega_1} p_n'(y)p_k(y)w(y)\mathrm{d}y$$

after changing our integration variable to y. Then, using integration by parts (with the substitution $u = p_k(y)w(y)$), we obtain

$$c_{n,k} = \int_{\Omega_1} p_n'(y)p_k(y)w(y)\mathrm{d}y = p_n(y)p_k(y)w(y)\big|_{\Omega_1}$$

$$- \int_{\Omega_1} p_n(y) \left(p_k(y)w'(y) + p_k'(y)w(y) \right) \mathrm{d}y.$$

Via (2.12) we have $w'(y) = -v'(y)w(y)$ and therefore

$$c_{n,k} = p_n(y)p_k(y)w(y)\big|_{\Omega_1} - \int_{\Omega_1} p_n(y) \left(p_k'(y) - p_k(y)v'(y) \right) w(y)\mathrm{d}y.$$

From Lemma 2.1, the integral with the $p_k'(y)$-term is zero. This finally gives us

$$c_{n,k} = p_n(y)p_k(y)w(y)\big|_{\Omega_1} + \int_{\Omega_1} p_n(y)p_k(y)v'(y)w(y)\mathrm{d}y.$$

From substituting this into (2.19) we have

$$p'_n(x) = w(y)p_n(y) \sum_{k=0}^{n-1} p_k(x)p_k(y) \Bigg|_{\Omega_1} + \int_{\Omega_1} p_n(y) \left(\sum_{k=0}^{n-1} p_k(x)p_k(y) \right) v'(y)w(y)dy.$$

Now notice, via Lemma 2.1, that the integral directly above is zero if $v'(y)$ is replaced by $v'(x)$. Thus, we can replace $v'(y)$ with $v'(y) - v'(x)$, in which case we see that

$$p'_n(x) = w(y)p_n(y) \sum_{k=0}^{n-1} p_k(x)p_k(y) \Bigg|_{\Omega_1}$$

$$+ \int_{\Omega_1} p_n(y) \left(\sum_{k=0}^{n-1} p_k(x)p_k(y) \right) (v'(y) - v'(x)) w(y)dy.$$

Next, via (2.15), we take $P_n(x) = \sqrt{\alpha_n}p_n(x)$ and use the Christoffel–Darboux Identity (Theorem 2.1) to evaluate each of the sums in the above equation. With some manipulations, we obtain (2.16).   □

Our next result relates $A(x)$ and $B(x)$ as in Theorem 2.2 and also plays a key role in our application to quantum mechanics.

Lemma 2.3. *The coefficients $A_n(x)$ and $B_n(x)$, as respectively defined in (2.17) and (2.18), satisfy the relation*

$$B_n(x) + B_{n+1}(x) = \frac{x - b_n}{a_n} A_n(x) - v'(x),$$

with $w(x)$ as defined in (2.12) and $w(x)$ vanishing at the boundary points of Ω_1.

Proof. Assuming that $w(x)$ vanishes at the boundary points of Ω_1, from (2.18) we have

$$B_n(x) + B_{n+1}(x) = a_n \int_{\Omega_1} \frac{v'(x) - v'(y)}{x - y} p_n(y)p_{n-1}(y)w(y)dy$$

$$+ a_{n+1} \int_{\Omega_1} \frac{v'(x) - v'(y)}{x - y} p_{n+1}(y)p_n(y)w(y)dy, \qquad (2.20)$$

which can be written as

$$B_n(x) + B_{n+1}(x) = \int_{\Omega_1} p_n(y) \frac{v'(x) - v'(y)}{x - y} (a_{n+1}p_{n+1}(y) + a_np_{n-1}(y)) w(y)dy. \qquad (2.21)$$

From (2.13), it follows that $a_{n+1}p_{n+1}(y) + a_np_{n-1}(y) = (y - b_n)p_n(y)$ and we therefore obtain

$$B_n(x) + B_{n+1}(x) = \int_{\Omega_1} \frac{v'(x) - v'(y)}{x - y}(y - b_n)p_n^2(y)w(y)\mathrm{d}y. \qquad (2.22)$$

We then take $y - b_n = (y - x) + (x - b_n)$ and see that the above integral becomes

$$(x - b_n)\int_{\Omega_1} \frac{v'(x) - v'(y)}{x - y}p_n^2(y)w(y)\mathrm{d}y - \int_{\Omega_1}(v'(x) - v'(y))p_n^2(y)w(y)\mathrm{d}y$$

$$= (x - b_n)\frac{A_n(x)}{a_n} + \int_{\Omega_1} v'(y)p_n^2(y)w(y)\mathrm{d}y - v'(x)\int_{\Omega_1} p_n^2(y)w(y)\mathrm{d}y$$

$$= \frac{x - b_n}{a_n}A_n(x) + \int_{\Omega_1} v'(y)p_n^2(y)w(y)\mathrm{d}y - v'(x),$$

where we have used (2.17) and the orthonormality of $\{p_n(x)\}_{n=0}^{\infty}$. From using integration by parts (with the substitution $u = v'(x)$) and orthonormality, we see that

$$\int_{\Omega_1} v'(y)p_n^2(y)w(y)\mathrm{d}y = v'(x)\big|_{\Omega_1} - \int_{\Omega_1} v''(y)\mathrm{d}y = 0.$$

Hence, we have

$$B_n(x) + B_{n+1}(x) = \frac{x - b_n}{a_n}A_n(x) - v'(x). \qquad \square$$

Theorem 2.3. *If $w(x)$ is as defined in (2.12) and $\{p_n(x)\}_{n=0}^{\infty}$ is an orthonormal set as defined by Definition 2.8, then $\{p_n(x)\}_{n=0}^{\infty}$ satisfies the (factored) second-order differential equation*

$$L_x^{2,n}\left(\frac{L_x^{1,n}p_n(x)}{A_n(x)}\right) = \frac{a_n}{a_{n-1}}A_{n-1}(x)p_n(x), \qquad (2.23)$$

with the differential operators $L_x^{1,n}$ and $L_x^{2,n}$ defined as

$$L_x^{1,n} := \frac{\mathrm{d}}{\mathrm{d}x} + B_n(x) \qquad (2.24)$$

$$L_x^{2,n} := -\frac{\mathrm{d}}{\mathrm{d}x} + B_n(x) + v'(x). \qquad (2.25)$$

Proof. In light of (2.24), we can write (2.16) as

$$L_x^{1,n}p_n(x) = A_n(x)p_{n-1}(x). \qquad (2.26)$$

Next, we define the weighted inner product

$$\langle p_m(x), p_n(x)\rangle_w := \int_{\Omega_1} p_m(x)p_n(x)w(x)\mathrm{d}x$$

and utilize the **Hilbert Space** where the $\langle p_n, p_n \rangle_w$ is finite and $p_n(x)\sqrt{w(x)}$ is zero at the end points of Ω_1 (finite or infinite). Thus, $L_x^{2,n} = \left(L_x^{1,n} \right)^*$.

Using (2.16) and (2.13) leads to the **adjoint** equation

$$\left(-\frac{\mathrm{d}}{\mathrm{d}x} + B_n(x) + v'(x) \right) p_{n-1}(x) = \frac{a_n}{a_{n-1}} A_{n-1}(x) p_n(x). \tag{2.27}$$

Thus, we can use (2.26) to readily obtain

$$\frac{L_x^{1,n} p_n(x)}{A_n(x)} = p_{n-1}(x)$$

and with (2.25) we can rewrite (2.27) as

$$L_x^{2,n} \left(\frac{L_x^{1,n} p_n(x)}{A_n(x)} \right) = \frac{a_n}{a_{n-1}} A_{n-1}(x) p_n(x)$$

and the theorem is established. $\square$

We additionally have an equivalent form for (2.23), which also emphasizes the fact that it is a second-order equation.

Corollary 2.1. *The differential equation (2.23) can equivalently be expressed as*

$$p_n''(x) + C_n(x) p_n'(x) + D_n(x) p_n(x) = 0, \tag{2.28}$$

where

$$C_n(x) := -v'(x) - \frac{A_n'(x)}{A_n(x)}, \tag{2.29}$$

$$D_n(x) := A_n(x) \frac{\mathrm{d}}{\mathrm{d}x} \left(\frac{B_n(x)}{A_n(x)} \right) - B_n(x) \left(v'(x) + B_n(x) \right) + \frac{a_n}{a_{n-1}} A_n(x) A_{n-1}(x). \tag{2.30}$$

Proof. By expanding the left-hand side of (2.23) and using some manipulations, we achieve (2.28). $\square$

Remark 2.2. We see that (2.24) is in fact a degree-lowering operator analogous to the ones used in Sheffer's analysis [22]. In contrast, we also note that (2.25) is a **degree-raising operator** or equivalently, a **ladder operator**. What has been shown through Corollary 2.1 is that every classical orthogonal polynomial sequence (written in orthonormal form) is a solution to a second-order linear differential equation or equivalently possesses a degree-raising operator.

In addition, classical orthogonal polynomials have an important connection to quantum mechanics via Theorem 2.3 (and Corollary 2.1). Namely, we can now show that the solution to the time-independent Schrödinger equation can be expressed in terms of $\{p_n(x)\}_{n=0}^\infty$ if (2.23), or equivalently (2.28), is satisfied.

Theorem 2.4. *The second-order linear differential equation* (2.23), *or equivalently* (2.28), *can be written in the* **Schrödinger form**:

$$\Psi_n''(x) + V(x;n)\Psi_n(x) = 0, \tag{2.31}$$

where

$$\Psi_n(x) := \frac{\exp(-v(x)/2)}{\sqrt{A_n(x)}} p_n(x) \tag{2.32}$$

and

$$V(x;n) = A_n(x)\frac{d}{dx}\left(\frac{B_n(x)}{A_n(x)}\right) - B_n(x)\left(v'(x) + B_n(x)\right) + \frac{a_n}{a_{n-1}}A_n(x)A_{n-1}(x)$$

$$+ \frac{1}{2}v''(x) + \frac{1}{2}\frac{d}{dx}\left(\frac{A_n'(x)}{A_n(x)}\right) - \frac{1}{4}\left(v'(x) + \frac{A_n'(x)}{A_n(x)}\right)^2. \tag{2.33}$$

Proof. Substituting $\Psi_n(x)$ and $V(x;n)$ into the left-hand side of (2.31) and using manipulations, we see that the resulting expression vanishes. □

In light of Theorem 2.4, we have the following discussion. For a particle confined to one dimension, the time-independent Schrödinger equation is written in the form

$$\psi''(x) + \frac{2m}{\hbar^2}(E - U(x))\psi(x) = 0, \tag{2.34}$$

where $\psi(x)$ represents the **wave function**, m the mass of a particle, $\hbar := h/(2\pi)$, where h is Plank's constant, E the total energy (constant), and $U(x)$ the potential energy. This equation is attributed to Erwin Schrödinger, who formulated it in late 1925 and published it in 1926, e.g., [18, 19]. We discuss this equation further.

To begin, the wave function $\psi(x)$ itself is not directly related to any actual physical phenomena. In addition, $\psi(x)$ may be a real- or complex-valued function. However, the square of its modulus (absolute magnitude) $|\psi(x)|^2$ when evaluated at a certain location in space is directly proportional to the *probability* of locating a particle in the same location. The quantity $|\psi(x)|^2$ is referred to as the **probability density**. In fact, from knowing $\psi(x)$ explicitly, the linear momentum, the energy, and other physical quantities of a particle can be inferred. We also mention that the entire field of quantum mechanics can basically be summarized as determining $\psi(x)$ for a given particle when its range of motion is restricted by potential fields.

The wave function $\psi(x)$ must adhere to certain physical restraints. One, since $|\psi(x)|^2$ is directly proportional to the probability, say $\mathcal{P}$, of locating a particle modeled by $\psi(x)$, it follows that the following must hold:

$$\int_{-\infty}^\infty |\psi(x)|^2 dx = \int_{-\infty}^\infty \mathcal{P}dx = 1.$$

Intuitively, this means that if the particle exists, it must be somewhere in space. The above relation is a normalization. The probability that the particle will be discovered in a certain region of space, say $[a,b] \subseteq \mathbb{R}$, is then given by

$$\mathcal{P}_{[a,b]} := \int_a^b |\psi(x)|^2 \mathrm{d}x \in [0,1].$$

Now that the wave function $\psi(x)$ is better understood, we can discuss our equation at hand. By letting $V(x) := E - U(x)$ we can of course equivalently write (2.34) as

$$\psi''(x) + \frac{2m}{\hbar} V(x)\psi(x) = 0.$$

From this relation, it is clear why (2.31) is named as such. Moreover, in essence, the time-independent Schrödinger equation describes how the wave function evolves over space. Our relation (2.34), and therefore (2.31), is useful in many physical situations when the potential energy of a particle does not depend upon time.

To complete our discussion, it is noteworthy to mention that the time-dependent Schrödinger equation can be written as

$$i\hbar \frac{\partial}{\partial t} \psi(x,t) = -\frac{\hbar^2}{2m} \frac{\partial^2}{\partial x^2} \psi(x,t) + U(x)\psi(x,t).$$

Furthermore, Schrödinger basically established his equation(s) based on thought experiments. In regard to his development, Richard Feynman said:

> Where did we get that equation from? Nowhere. It is not possible to derive it from anything you know. It came out of the mind of Schrödinger.

For more details related to the Schrödinger equations and quantum mechanics, refer to [6, 20].

Next, we have another relationship between $A_n(x)$ and $B_n(x)$, which we use specifically in Example 2.1.

Theorem 2.5. *The coefficients $A_n(x)$ and $B_n(x)$ as respectively in (2.17) and (2.18) satisfy*

$$B_{n+1}(x) - B_n(x) = \frac{a_{n+1}A_{n+1}(x)}{x - b_n} - \frac{a_n^2 A_{n-1}(x)}{a_{n-1}(x - b_n)} - \frac{1}{x - b_n}. \qquad (2.35)$$

Proof. Refer to [14]. □

We note that (2.35) is referred to as the **string equation**.

We use the following fundamental **special function** in what follows.

Definition 2.9. The **gamma function**, $\Gamma(z)$, is defined as

$$\Gamma(z) := \int_0^\infty \tau^{z-1} \mathrm{e}^{-\tau} \mathrm{d}\tau, \quad Re(z) > 0.$$

Also, $\Gamma(z)$ satisfies the functional equation

$$\Gamma(z+1) = z\Gamma(z), \tag{2.36}$$

which in fact extends $\Gamma(z)$ to a meromorphic function with poles at all of the non-positive integers. See Chap. 2 of [17] for an extensive development of the gamma function.

From (2.36) and (2.6), we see that

$$(c)_n = \frac{\Gamma(c+n)}{\Gamma(c)}. \tag{2.37}$$

Example 2.5. We now discuss the details of how a particular Sheffer *A-Type 0* orthogonal set solves the Schrödinger Form (2.31). We first note that each classical orthogonal set yields a unique set of functions $\{A_n(x), B_n(x)\}$, which are of course dependent on the polynomials themselves, the corresponding interval of orthogonality Ω_1, the weight function $w(x)$ [and therefore $v(x)$ as in (2.12)] as well as the recursion coefficient a_n. Therefore, each of the three Sheffer *A-Type 0* sets with continuous orthogonality relations (Hermite, Laguerre and Meixner–Pollaczek) has a unique wave function $\Psi_n(x)$ and consequently a unique kinetic energy function $V(x;n)$.

In general, we must first write our orthogonal set in orthonormal form. Then we must find a_n, $v(x)$, $A_n(x)$, and $B_n(x)$. For our particular example, we use the Laguerre polynomials, which are defined as

$$L_n^{(\alpha)}(x) := \frac{(\alpha+1)_n}{n!} \sum_{k=0}^{n} \frac{(-n)_k}{(\alpha+1)_k} \frac{x^k}{k!}, \quad \alpha > -1 \tag{2.38}$$

and have the following (continuous) orthogonality relation

$$\int_0^{\infty} L_m^{(\alpha)}(x) L_n^{(\alpha)}(x) x^{\alpha} e^{-x} dx = \frac{\Gamma(\alpha+1+n)}{n!} \delta_{m,n}.$$

The restriction on α is essential, as $\{L_n^{(\alpha)}(x)\}_{n=0}^{\infty}$ is undefined for all $\alpha \le -1$. It is worth noting that taking into account that $(\alpha+1)_n/(\alpha+1)_k = (\alpha+k+1)_{n-k}$ leads to the equivalent form

$$L_n^{(\alpha)}(x) = \frac{1}{n!} \sum_{k=0}^{n} \frac{(-n)_k}{k!} (\alpha+k+1)_{n-k} x^k.$$

Moreover, using this relationship, we can define the Laguerre polynomials for all $\alpha \in \mathbb{R}$.

For this application, we first assume $\alpha > 0$. We begin by putting our orthogonality relation in orthonormal form using (2.37) on $\Gamma(\alpha+1+n)$:

$$\int_0^\infty L_m^{(\alpha)}(x) L_n^{(\alpha)}(x) \frac{n!}{(\alpha+1)_n} \frac{x^\alpha e^{-x}}{\Gamma(\alpha+1)} dx = \delta_{m,n}. \tag{2.39}$$

This enables us to define

$$p_n(x) = (-1)^n \sqrt{\frac{n!}{(\alpha+1)_n}} L_n^{(\alpha)}(x) \tag{2.40}$$

and

$$w(x) = \frac{x^\alpha e^{-x}}{\Gamma(\alpha+1)}. \tag{2.41}$$

We first determine $A_n(x)$ as in (2.17) and begin by evaluating the boundary conditions, noting that the interval of orthogonality for the Laguerre polynomials is $\Omega_1 = [0, \infty)$. We readily obtain

$$\lim_{b \to \infty} \frac{p_n^2(y) w(y)}{y-x} \bigg|_{y=0}^{y=b} = \frac{n!}{\Gamma(\alpha+1)(\alpha+1)_n} \lim_{b \to \infty} \frac{b^\alpha \left[L_n^{(\alpha)}(b)\right]^2}{e^b(b-x)} - 0 = 0.$$

Next, we see from (2.41) that

$$w(x) = \exp\left(-\ln\left(\frac{e^x \Gamma(\alpha+1)}{x^\alpha}\right)\right)$$

and therefore

$$v(x) = \ln\left(\frac{e^x \Gamma(\alpha+1)}{x^\alpha}\right) \tag{2.42}$$

$$\Rightarrow v'(x) = \frac{x-\alpha}{x}.$$

Moreover,

$$\frac{v'(x) - v'(y)}{x-y} = \frac{\alpha}{xy}.$$

Putting all of this together, we thus far have

$$A_n(x) = \alpha \frac{a_n}{x} \int_0^\infty p_n^2(y) \frac{y^{\alpha-1} e^{-y}}{\Gamma(\alpha+1)} dy.$$

We evaluate the above integral using integration by parts (with the substitution $u = p_n^2(y) e^{-y}$), which leads to

$$A_n(x) = \frac{a_n}{x}\left(\lim_{b\to\infty}\frac{p_n^2(y)y^\alpha}{e^y}\bigg|_0^b + \int_0^\infty p_n^2(y)\frac{y^\alpha e^{-y}}{\Gamma(\alpha+1)}dy\right.$$

$$\left. -2\int_0^\infty p_n(y)p_n'(y)\frac{y^\alpha e^{-y}}{\Gamma(\alpha+1)}dy\right).$$

In the result directly above, it is clear that the boundary conditions equal zero. The middle integral is equal to 1 via the orthonormality relation in Definition 2.8. Using the fact that $\deg(p_n'(y)) = n-1$ and Lemma 2.1, it follows that the last integral is equal to zero. This results in

$$A_n(x) = \frac{a_n}{x}. \tag{2.43}$$

We derive $B_n(x)$ in a similar fashion. In this case, it is also immediate that the boundary conditions for (2.18) are zero as well. Therefore, we have the following

$$B_n(x) = \alpha\frac{a_n}{x}\int_0^\infty p_n(y)p_{n-1}(y)\frac{y^{\alpha-1}e^{-y}}{\Gamma(\alpha+1)}dy.$$

We use integration by parts on the above integral (with the substitution $u = p_n(y)p_{n-1}(y)e^{-y}$) and again call upon Lemma 2.1 to achieve

$$B_n(x) = -\frac{a_n}{x}\int_0^\infty p_{n-1}(y)p_n'(y)\frac{y^\alpha e^{-y}}{\Gamma(\alpha+1)}dy. \tag{2.44}$$

In order to fully evaluate (2.44), we momentarily digress. We assume that $\{\mathcal{Q}_n(x)\}_{n=0}^\infty$ defines a generic orthogonal set, with $\mathcal{Q}_n(x) = q_{n,n}x^n + \mathcal{O}(x^{n-1})$. Then, using Lemma 2.1, we know that

$$\int_\Omega \mathcal{Q}_n^2(y)w(y)dy = \int_\Omega \mathcal{Q}_n(y)[q_{n,n}y^n + \mathcal{O}(y^{n-1})]w(y)dx = q_{n,n}\int_\Omega \mathcal{Q}_n(y)y^n w(y)dy.$$

Also, from directly evaluating (2.13), it is clear that the coefficient of y^n in $p_n(y)$ is $(a_1\cdots a_n)^{-1}$. Thus, using this fact, our simple relation directly above and orthonormality, we have

$$a_n\int_0^\infty p_{n-1}(y)p_n'(y)\frac{y^\alpha e^{-y}}{\Gamma(\alpha+1)}dy = a_n\int_0^\infty p_{n-1}(y)\left[\frac{ny^{n-1}}{a_1\cdots a_n}\right]\frac{y^\alpha e^{-y}}{\Gamma(\alpha+1)}dy$$

$$= n\int_0^\infty p_{n-1}(y)\left[\frac{y^{n-1}}{a_1\cdots a_{n-1}}\right]\frac{y^\alpha e^{-y}}{\Gamma(\alpha+1)}dy$$

$$= n\int_0^\infty p_{n-1}^2(y)\frac{y^\alpha e^{-y}}{\Gamma(\alpha+1)}dy$$

$$= n$$

and we conclude that

$$B_n(x) = \frac{-n}{x}. \tag{2.45}$$

To get a complete expression for $A_n(x)$, we must determine the recursion coefficient a_n, which of course can be done by using (2.14) and taking into account that the monic three-term recurrence relation for the Laguerre polynomials is

$$xQ_n(x) = Q_{n+1}(x) + (2n + \alpha + 1)Q_n(x) + n(n + \alpha)Q_{n-1}(x),$$

$$Q_n(x) := (-1)^n n! L_n^{(\alpha)}(x).$$

As it turns out, from our analysis above, we can also obtain a_n. We first substitute our relations (2.43) and (2.45) into the string equation (2.35), which yields

$$a_{n+1}^2 - a_n^2 = b_n. \tag{2.46}$$

Using Lemma 2.3 in the same way, we get

$$b_n = \alpha + 2n + 1. \tag{2.47}$$

Hence, (2.46) and (2.47) imply that

$$a_n = \sqrt{n(n + \alpha)}, \tag{2.48}$$

which follows since $n(n + \alpha)$ is nonnegative. Therefore, we have

$$A_n(x) = \frac{\sqrt{n(n + \alpha)}}{x}. \tag{2.49}$$

Furthermore, we can disregard the restriction $\alpha > 0$, as the above relation is clearly valid for $\alpha > -1$.

To evaluate the wave function $\Psi_n(x)$ in (2.32), we first note that

$$\exp(-v(x)/2) = \frac{x^{\alpha/2}}{e^{x/2}(\Gamma(\alpha + 1))^{1/2}}.$$

Using this relation and (2.49) we have

$$\Psi_n(x) = \frac{x^{\frac{\alpha+1}{2}} e^{-x/2} p_n(x)}{n^{1/4}(n + \alpha)^{1/4}(\Gamma(\alpha + 1))^{1/2}}. \tag{2.50}$$

Now, we substitute (2.49), (2.45), (2.48) and (2.42) into (2.33) and use manipulations to obtain the kinetic energy function

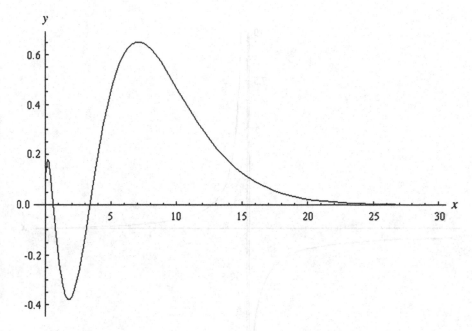

Fig. 2.1 $\Psi_2(x)$ for $\alpha = 0$

$$V(x;n) = \frac{-x^2 + 2(\alpha + 2n + 1)x + 1 - \alpha^2}{4x^2}. \tag{2.51}$$

Lastly, as a concrete example we take $\alpha = 0$ and $n = 2$, which gives the following wave and kinetic energy functions:

$$\Psi_2(x) = \frac{1}{2}\sqrt{\frac{xe^{-x}}{2}}p_2(x) \quad \text{and} \quad V(x;2) = \frac{-x^2 + 10x + 1}{4x^2}.$$

The graph of $\Psi_2(x)$ is displayed in Fig. 2.1.

We also display the graph of the kinetic energy function $V(x;2)$ in Fig. 2.2.

For both of these plots we used Mathematica®.

Example 2.6. The motivated reader can show that for the Hermite polynomials, $v(x) = x^2$, $A_n(x) = \sqrt{2n}$ and $B_n(x) = 0$ and determine $\Psi_n(x)$ and $V(x;n)$ accordingly.

Next, we show how (2.16) of Theorem 2.2 and (2.28) of Corollary 2.1 can be applied to a specific Sheffer *A-Type 0* set.

Example 2.7. Now that we have $A_n(x)$ and $B_n(x)$ for the Laguerre polynomials via Example 2.5, we can utilize these results in determining the specific differential equations (2.16) and (2.23) [and therefore (2.28)] that are satisfied by this particular *A-Type 0* orthogonal set. Using $p_n(x)$ in (2.40), as well as (2.17) and (2.18) for $A_n(x)$ and $B_n(x)$, respectively, we see that (2.16) becomes

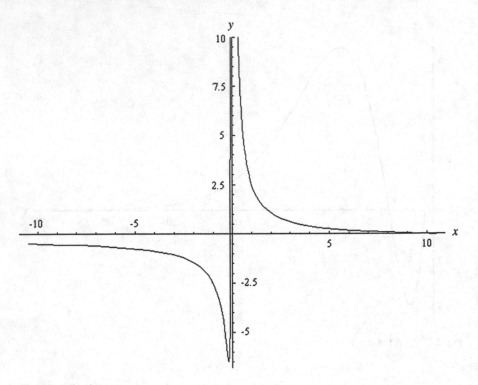

Fig. 2.2 $V(x;2)$ for $\alpha = 0$

$$\frac{d}{dx}L_n^{(\alpha)}(x) = \frac{n}{x}L_n^{(\alpha)}(x) - \frac{\alpha+n}{x}L_{n-1}^{(\alpha)}(x),$$

since $(\alpha+1)_n/(\alpha+1)_{n-1} = \alpha+n$. Similarly, (2.23) and (2.28) take on the form

$$x\frac{d^2}{dx^2}L_n^{(\alpha)}(x) + (1+\alpha+x)\frac{d}{dx}L_n^{(\alpha)}(x) + nL_n^{(\alpha)}(x) = 0.$$

2.4   Difference Equations

In Sects. 2.2 and 2.3, we gave examples of how continuous orthogonal polynomials are important in differential equations. Here, we show how discrete orthogonal polynomials play an important role in *difference equations*. In particular, what follows is a discrete analogue of the characterizations of Sect. 2.3 and is essentially based on [13].

To begin, we assume that $\{p_n(x)\}_{n=0}^{\infty}$ satisfies a discrete orthogonality relation as in (2.2). We expand on this assumption for this section by assuming the discrete weight function $w(x)$ is supported on the set $\{s, s+1, \ldots, t\} \subset \mathbb{R}$, where s is a finite number, but t may be either finite or infinite. We then write the orthogonality relation (2.2) as

$$\sum_{j=s}^{t} p_m(j)p_n(j)w(j) = \beta_n \delta_{m,n}, \quad w(s-1) = w(t+1) = 0. \tag{2.52}$$

In order to develop our analogues, we must find a discrete version of $v(x)$ as in (2.12). We call this analogue $u(x)$, which is defined as

$$u(x) := \frac{w(x-1) - w(x)}{w(x)}. \tag{2.53}$$

The discrete analogue of the Christoffel–Darboux Identity (Theorem 2.1) is

$$\sum_{k=0}^{n-1} \frac{p_k(x)p_k(y)}{\beta_n} = \frac{\gamma_{n-1}}{\gamma_n \beta_n} \left(\frac{p_n(x)p_{n-1}(y) - p_n(y)p_{n-1}(x)}{x-y} \right), \tag{2.54}$$

where γ_n denotes the leading coefficient of $p_n(x)$—we use this notation throughout this entire section.

We now state the analogue of Theorem 2.2.

Theorem 2.6. Let $\{p_n(x)\}_{n=0}^{\infty}$ satisfy (2.52). Then, we have

$$\Delta p_n(x) = -B_n(x)p_n(x) + A_n(x)p_{n-1}(x), \tag{2.55}$$

where the linear difference operator Δ is defined as

$$\Delta f(x) := f(x+1) - f(x),$$

with

$$A_n(x) = \frac{\gamma_{n-1}}{\gamma_n \beta_{n-1}} \left(\frac{p_n(t+1)p_n(t)}{t-x} w(t) + \sum_{j=s}^{t} p_n(j)p_n(j-1) \frac{u(x+1) - u(j)}{x+1-j} w(j) \right) \tag{2.56}$$

and

$$B_n(x) = \frac{\gamma_{n-1}}{\gamma_n \beta_{n-1}} \left(\frac{p_n(t+1)p_{n-1}(t)}{t-x} w(t) \right.$$

$$\left. + \sum_{j=s}^{t} p_n(j)p_{n-1}(j-1) \frac{u(x+1) - u(j)}{x+1-j} w(j) \right). \tag{2.57}$$

The s and t-values are as in (2.52).

Proof. The proof is similar to that of Theorem 2.2 and is left to the reader as an exercise. $\square$

Notice that in Theorem 2.6, $\{p_n(x)\}_{n=0}^{\infty}$ was only assumed to be an orthogonal set with respect to the discrete weight function $w(x)$. If it is further assumed that $\{p_n(x)\}_{n=0}^{\infty}$ is an orthonormal set, i.e., $\beta_n \equiv 1$, then it satisfies (2.13). Consequently, $\gamma_{n-1}/\gamma_n = a_n$, and $A_n(x)$ and $B_n(x)$ as respectively in (2.56) and (2.57) take on the form

$$A_n(x) = a_n \left(\frac{p_n(t+1)p_n(t)}{t-x} w(t) + \sum_{j=s}^{t} p_n(j)p_n(j-1) \frac{u(x+1)-u(j)}{x+1-j} w(j) \right) \quad (2.58)$$

and

$$B_n(x) = a_n \left(\frac{p_n(t+1)p_{n-1}(t)}{t-x} w(t) + \sum_{j=s}^{t} p_n(j)p_{n-1}(j-1) \frac{u(x+1)-u(j)}{x+1-j} w(j) \right).$$

$$(2.59)$$

To demonstrate the applicability of (2.55) to the discrete Sheffer *A-Type 0* orthogonal sets, we consider the Meixner polynomials in our next example. To complete this example, we need the very useful **Chu-Vandermonde Sum**, which is

$$_2F_1 \left(\begin{array}{c} -n,b \\ c \end{array} \middle| 1 \right) = \frac{(c-b)_n}{(c)_n}.$$

It is worthwhile to note that the Chu-Vandermonde Sum can be obtained as the terminating version of the **Gauss Sum**:

$$_2F_1 \left(\begin{array}{c} a,b \\ c \end{array} \middle| 1 \right) = \frac{\Gamma(c)\Gamma(c-a-b)}{\Gamma(c-a)\Gamma(c-b)}, \quad \mathrm{Re}(c-a-b) > 0.$$

Namely, if in the Gauss Sum we replace a with $-n$, we are left with the Chu-Vandermonde Sum. Refer to Chap. 4 of [17] and Chap. 1 of [25] for more information on these and other similar sums.

Example 2.8. Here, we sketch the details of deriving the coefficients $A_n(x)$ and $B_n(x)$ in (2.55) for the Meixner polynomials. These polynomials can be defined as

$$M_n(x;\beta,c) := {_2F_1} \left(\begin{array}{c} -n,-x \\ \beta \end{array} \middle| 1-\frac{1}{c} \right), \quad (2.60)$$

where $x = 0,1,2,\ldots$. The weight function for the Meixner polynomials is

$$w(x) = \frac{(\beta)_x c^x}{x!}, \quad x = 0,1,2,\ldots \quad (2.61)$$

via the discrete orthogonality relation

$$\sum_{x=0}^{\infty} M_m(x;\beta;c)M_n(x;\beta;c)\frac{(\beta)_x c^x}{x!} = \frac{n!(1-c)^{-\beta}}{c^n(\beta)_n}\delta_{m,n}. \tag{2.62}$$

Then, it follows from (2.53) that

$$u(x) = \frac{x}{(\beta+x-1)c} - 1.$$

Next, we consider $q(x)$ such that $\deg(q(x)) \leq n$ and take c to be an arbitrary constant. Then, from our discrete analogue of the Christoffel–Darboux Identity (2.54), we observe that

$$\sum_{j=s}^{t} \frac{p_n(j)q(j)}{j-c}w(j) = q(c)\sum_{j=s}^{t}\frac{p_n(j)}{j-c}w(j), \tag{2.63}$$

which follows from (2.52). Using (2.63), as well as $A_n(x)$ in (2.56) and the Meixner weight function (2.61), it can be shown after several manipulations that

$$A_n(x) = \frac{\gamma_{n-1}}{\gamma_n\beta_{n-1}}\frac{p_n(-\beta)}{(\beta+x)c}\sum_{k=0}^{\infty}\frac{(\beta-1)_k c^k}{k!}p_n(k).$$

Using the definition of the Meixner polynomials (2.60), it can also be shown further that

$$A_n(x) = \frac{\gamma_{n-1}}{\gamma_n\beta_{n-1}}\frac{p_n(-\beta)}{(\beta+x)c}\sum_{j=0}^{n}\frac{(-n)_j}{j!(\beta)_j}\left(1-\frac{1}{c}\right)^j(-1)^j\sum_{k=j}^{\infty}\frac{(\beta-1)_k c^k}{(k-j)!}.$$

For the next step, we call upon the **binomial theorem**, which we write using the Pochhammer symbol:

$$\sum_{n=0}^{\infty}\frac{(a)_n}{n!}z^n = (1-z)^{-a},$$

where, in general, we must require $|z| < 1$ if a is not a negative integer. We apply the binomial theorem to the right most sum in our last manipulation for $A_n(x)$ and also utilize the Chu-Vandermonde Sum to obtain

$$A_n(x) = \frac{\gamma_{n-1}}{\gamma_n\beta_{n-1}}\frac{p_n(-\beta)(1-c)^{1-\beta}n!}{c(\beta+x)(\beta)_n}.$$

The binomial theorem and (2.60) lead to

$$p_n(-\beta) = {}_1F_0\left(\begin{matrix}-n\\-\end{matrix}\bigg|1-\frac{1}{c}\right) = \frac{1}{c^n}. \tag{2.64}$$

Then, from (2.60) and (2.62), we obtain

$$\gamma_n = \frac{1}{(\beta)_n}\left(1 - \frac{1}{c}\right)^n \quad \text{and} \quad \beta_n = \frac{n!}{c^n(1-c)^\beta(\beta)_n}.$$

By using (2.64) and the expressions for γ_n and β_n above to find γ_{n-1}/γ_n and β_{n-1}, we eventually have

$$A_n(x) = \frac{-n}{c(\beta + x)}.$$

To derive the expression for $B_n(x)$, we use (2.61), (2.56), (2.57), (2.63) and (2.64) to arrive at

$$B_n(x) = \frac{p_{n-1}(-\beta)}{p_n(-\beta)}A_n(x) = \frac{-n}{\beta + x}.$$

Collecting all of this analysis, we establish the following theorem.

Theorem 2.7. *With respect to the operator Δ as defined in Theorem 2.6, the Meixner polynomials satisfy the first-order difference equation*

$$\Delta M_n(x;\beta,c) = \left(\frac{n}{\beta + x}\right)M_n(x;\beta,c) - \left(\frac{n}{c(\beta + x)}\right)M_{n-1}(x;\beta,c).$$

Proof. See all of the above analysis.   □

2.5   Gaussian Quadrature

Informally, the idea behind *Gaussian Quadrature* is to approximate the integral of a given function, say $f(x)$, multiplied by a weight function, with a linear combination of $f(x)$ at known x-values. That is given $f(x)$, we want to obtain

$$\int_\mathbb{R} f(x)w(x)\mathrm{d}x \approx c_1 f(x_1) + c_2 f(x_2) + \cdots + c_N f(x_N).$$

Of course, one would need to develop a method to accomplish this, i.e., how to specifically determine the c_i-terms, the N-value, and the set $\{x_k\}_{k=1}^N$. The essence of such a method was essentially devised by Carl F. Gauss.

As we have mentioned, this introduction is of course very informal, but should give a clear idea of our motivation. In this section, we show how Gaussian Quadrature is connected to classical orthogonal polynomial sequences and give a specific example using the Hermite *A-Type 0* polynomials. To begin, we must first discuss two important ideas that are necessary in the construction and proof of the main theorem of this section: the Lagrange Interpolation Polynomial and the nature of the zeros of classical monic orthogonal polynomials.

Definition 2.10. For a set of distinct points $\{x_1,\ldots,x_n\}$, called *nodes*, the *Lagrange Fundamental Polynomial*, which we denote $l_k(x)$, is

$$l_k(x) := \prod_{j=1,j\neq k}^{n} \frac{(x-x_j)}{(x_k-x_j)} = \frac{S_n(x)}{S'_n(x_k)(x-x_k)}, \quad k \in \{1,\ldots,n\},$$

with

$$S_n(x) = \prod_{j=1}^{n}(x-x_j).$$

As an immediate consequence, the *Lagrange Interpolation Polynomial* of a function $f(x)$ at the nodes $\{x_1,\ldots,x_n\}$ is the unique polynomial $L(x)$ such that $\deg(L(x)) = n - 1$ and $f(x_j) = L(x_j)$ is satisfied. In addition, it is also immediate that $L(x)$ can be written as

$$L(x) = \sum_{k=1}^{n} l_k(x)f(x_k) = \sum_{k=1}^{n} f(x_k)\frac{S_n(x)}{S'_n(x_k)(x-x_k)}. \tag{2.65}$$

We emphasize that Lagrange Interpolation Polynomials do *not* form an orthogonal set and are not of any Sheffer *Type*. Nonetheless, as we discussed, they are necessary in the construction of the main result of this section. The interested reader can however refer to [24] for a more in-depth coverage of Lagrange Interpolation, including convergence properties.

We next state an important result about the zeros of classical monic orthogonal polynomials that will be needed in our main theorem. Moreover, for establishing this theorem, we call upon the Christoffel–Darboux Identity and its limiting case, which makes for a quite simple proof. For the remainder of this section we assume that $\{P_n(x)\}_{n=0}^{\infty}$ is a monic sequence of orthogonal polynomials that satisfy (2.4).

Theorem 2.8. *If* $\mathcal{S} := \{P_n(x)\}_{n=0}^{\infty}$ *satisfies the monic three-term recurrence relation* (2.4), *then the zeros of each* $P_k(x) \in \mathcal{S}$ *are both real and simple (distinct).*

Proof. Assume that $P_k(x)$ has a complex zero, say z. Then, since complex zeros of polynomials with real coefficients come in conjugate pairs, we know that $\bar{z}$ is also a zero of $P_k(x)$. Now let $x = z$ and $y = \bar{z}$ and substitute these into (2.8). By the properties of the complex conjugate, we know that $P_k(\bar{z}) = \overline{P_k(z)}$ and therefore the right-hand side of (2.8) is zero. However, the left-hand side is nonzero. This contradiction implies that the zeros of each $P_k(x) \in \mathcal{S}$ are all real.

Next, assume that $P_k(x)$ has a zero of multiplicity 2, say x_0. Then, the right-hand side of (2.11) is clearly zero since $P_k(x_0) = P'_k(x_0) = 0$, while the left-hand side is nonzero (positive). Therefore, we know that the zeros of each $P_k(x) \in \mathcal{S}$ are simple. $\square$

We now can prove the following.

Theorem 2.9 (Gauss–Jacobi Mechanical Quadrature). *Let* $\{P_n(x)\}_{n=0}^{\infty}$ *satisfy the monic three-term recurrence relation* (2.4), *with real and simple zeros ordered as*

$$x_{N,1} > x_{N,2} > \cdots > x_{N,N}.$$

Then, given a positive integer N, there exist a unique sequence of positive numbers $\{\lambda_k\}_{k=1}^{N}$ such that

$$\int_{\mathbb{R}} p(x)w(x)\mathrm{d}x = \sum_{k=1}^{N} \lambda_k p(x_{N,k}), \tag{2.66}$$

*which is valid for all polynomials $p(x)$ such that $\deg(p(x)) \leq 2N - 1$. Moreover, the λ_k-values, called the **Christoffel Numbers**, are not dependent on the polynomial $p(x)$ and can be computed via*

$$\lambda_k = \int_{\mathbb{R}} \frac{P_N(x)w(x)\mathrm{d}x}{P_N'(x_{N,k})(x - x_{N,k})}. \tag{2.67}$$

Proof. Let $L(x)$ be the Lagrange Interpolation Polynomial in (2.65) of $p(x)$ at the nodes $\{x_{N,k}\}_{k=1}^{N}$, which we assign to be the zeros of $\{P_n(x)\}_{n=0}^{\infty}$. These choices are permissible, as each of the zeros $\{P_n(x)\}_{n=0}^{\infty}$ is real and simple via Theorem 2.8 and can be ordered as in the statement of this theorem. Therefore,

$$L(x) = p(x) \quad \text{at } x = x_{N,k}, \ \forall k \in \{1,2,\ldots,N\}.$$

Thus, there exists a polynomial $q(x)$ with $\deg(q(x)) \leq N - 1$ such that

$$p(x) - L(x) = P_N(x)q(x).$$

Multiplying the result directly above by $w(x)$ and integrating gives

$$\int_{\mathbb{R}} p(x)w(x)\mathrm{d}x = \int_{\mathbb{R}} L(x)w(x)\mathrm{d}x + \int_{\mathbb{R}} P_N(x)q(x)w(x)\mathrm{d}x.$$

From Lemma 2.1, the rightmost integral is zero. By (2.65) with $S(x)$ replaced by $P_N(x)$ we see that our relation above becomes

$$\int_{\mathbb{R}} p(x)w(x)\mathrm{d}x = \sum_{k=1}^{N} \left(\int_{\mathbb{R}} \frac{P_N(x)w(x)\mathrm{d}x}{P_N'(x_{N,k})(x - x_{N,k})} \right) p(x_{N,k})$$

and (2.66) and (2.67) are both satisfied. Since we have proven (2.66), we can now apply it to $p(x) = P_N^2(x)/(x - x_{N,k})^2$, from which it follows that

$$\lambda_k = \int_{\mathbb{R}} \left(\frac{P_N(x)}{P_N'(x_{N,k})(x - x_{N,k})} \right)^2 w(x)\mathrm{d}x.$$

This implies that $\{\lambda_k\}_{k=1}^{N}$ is a unique sequence of positive numbers. $\qquad\square$

We now demonstrate Theorem 2.9 concretely.

Example 2.9. We apply Theorem 2.9 to the polynomial $p(x) := 3x^3 - 2x^2$ using the Hermite polynomials, in which case $\Omega_1 = \mathbb{R}$. Since the weight function for the Hermite polynomials is $w(x) = e^{-x^2}$, we wish to evaluate

$$\int_{\mathbb{R}} \left(3x^3 - 2x^2\right) e^{-x^2} dx$$

via the right-hand side of (2.66). The Hermite polynomials are defined as

$$H_n(x) := n! 2^n \sum_{k=0}^{\lfloor n/2 \rfloor} \frac{(-1)^k x^{n-2k}}{2^{2k} k! (n-2k)!},$$

where $\lfloor n/2 \rfloor$ is the **floor function**. Through some manipulation, we can write these polynomials in the following hypergeometric form as in (2.5):

$$H_n(x) := (2x)^n {}_2F_0 \left(\begin{matrix} -n/2, (1-n)/2 \\ - \end{matrix} \middle| -\frac{1}{x^2} \right)$$

from which it is clear that the leading coefficient of $H_n(x)$ is 2^n. So, the monic form of the Hermite polynomials is $P_n(x) = 2^{-n} H_n(x)$. We next take $N = 3$ and thus

$$P_3(x) = \frac{1}{8} H_3(x) = x^3 - \frac{3}{2}x \quad \Rightarrow \quad P_3'(x) = 3x^2 - \frac{3}{2}.$$

The zeros of $P_3(x)$ give us the nodes

$$x_{3,1} = \sqrt{3/2},$$

$$x_{3,2} = 0,$$

$$x_{3,3} = -\sqrt{3/2}.$$

Also, we evaluate P and P' at each of these nodes to respectively obtain

$$P(x_{3,1}) = \frac{3}{4} \left(3\sqrt{6} - 4\right)$$

$$P(x_{3,2}) = 0$$

$$P(x_{3,3}) = -\frac{3}{4} \left(3\sqrt{6} - 4\right)$$

and

$$P'(x_{3,1}) = 3,$$

$$P'(x_{3,2}) = -3/2,$$
$$P'(x_{3,3}) = 3.$$

We next compute the λ_k-values using

$$\lambda_k = \int_{\mathbb{R}} \frac{P_3(x)e^{-x^2}\,dx}{P_3'(x_{3,k})(x - x_{3,k})}, \quad k = 1, 2, 3.$$

Specifically, this gives

$$\lambda_1 = \sqrt{\pi}/6,$$
$$\lambda_2 = 2\sqrt{\pi}/3,$$
$$\lambda_3 = \sqrt{\pi}/6.$$

Putting all of this together, after some calculations we obtain

$$\sum_{k=1}^{3} \left(\int_{\mathbb{R}} \frac{P_3(x)e^{-x^2}\,dx}{P_3'(x_{3,k})(x - x_{3,k})} \right) p(x_{3,k}) = -\sqrt{\pi}$$

and of course

$$\int_{\mathbb{R}} \left(3x^3 - 2x^2 \right) e^{-x^2}\,dx = -2 \int_{\mathbb{R}} x^2 e^{-x^2}\,dx = -2 \left(\frac{\sqrt{\pi}}{2} \right) = -\sqrt{\pi}$$

via the gamma function in Definition 2.9.

Also, the interested reader can consider instead using the Meixner–Pollaczek or the Laguerre polynomials for this example.

To summarize, we have established a numerical estimate of an integral by picking *optimal* abscissas (the x-values) at which to evaluate the function. The Gauss–Jacobi Mechanical Quadrature Theorem (Theorem 2.9) states that these values are exactly the roots of the orthogonal polynomial for the same interval and weight function as the integral being approximated. Gaussian Quadrature is optimal since it fits all polynomials up to degree $2N - 1$.

The natural question is of course how Gaussian Quadrature applies when $f(x)$ is a function more general than a polynomial of degree at most $2N - 1$. A detailed analysis of such a development and related results is quite detailed and a discussion of this nature is not germane to the work at hand. Nonetheless, it is certainly worthwhile to cover one such fundamental result and we refer the interested reader to [24] for a thorough treatment.

The result below essentially states that for large N, the quadrature of Theorem 2.9 becomes very close to the actual value of the integral $\int_a^b f(x)\,dx$, for which the integral exists.

Theorem 2.10. *Let $w(x) > 0$ be a weight function defined on an arbitrary interval $[a,b]$ and $Q_N(f(x)) := \sum_{k=1}^{N} \lambda_k p(x_{N,k})$ the corresponding Gauss–Jacobi Mechanical Quadrature as in Theorem 2.9, where the λ_k-values are the Christoffel Numbers (2.67). Then the* **quadrature convergence**

$$\lim_{N \to \infty} Q_N(f(x)) - \int_a^b f(x)\mathrm{d}x = 0$$

holds for an arbitrary function $f(x)$ for which the above integral exists.

Proof. Refer to Chap. 15 of [24]. □

We conclude this section by briefly addressing two additional types of similar quadrature formulas. The first is **Radau Quadrature** [10], which is also a Gaussian Quadrature-like formula. In essence, Radau Quadrature requires $N+1$ points and fits all polynomials up to degree $2N$ and hence yields exact results for all polynomials of degree $2N - 1$. This procedure requires the use of a weight function $W(x)$ in which the endpoint -1 in the interval $[-1,1]$ is included in a total of N abscissas, leading to $N - 1$ free variables.

Also, **Laguerre-Gauss Quadrature** (often also called Gauss-Laguerre Quadrature) [2, 8] is additionally a Gaussian-like quadrature over the interval $[0, \infty)$ with the Laguerre $(L_n^{(0)}(x))$ weight function $W(x) = \mathrm{e}^{-x^2}$. This procedure also fits all polynomials up to degree $2N - 1$.

2.6 Problems

We finalize this section by stating some interesting problems that naturally arise from the analysis we have covered.

Problem 1. Supply the rigorous details of Meixner–Pollaczek case of Example 2.4 and also consider applying the inverse method to the remaining three *A-Type 0* sets. For the Hermite polynomials $c_n := 1/n!$, for the Meixner polynomials $c_n := (\beta)_n/n!$, and for the Krawtchouk polynomials $c_n := C(N,n)$.

Problem 2. Develop an analogue of Example 2.5 for the Meixner–Pollaczek polynomials.

Problem 3. Construct an analogue of Example 2.8 for the other Sheffer Sequences that satisfy a discrete orthogonality, i.e., the Charlier and Krawtchouk polynomials.

Problem 4. Establish a discrete analogue of Theorem 2.4 and apply it to the Meixner, Charlier and Krawtchouk polynomials.

References

1. W.A. Al-Salam, *Characterization theorems for orthogonal polynomials*, in: Nevai, P. (Ed.), Orthogonal Polynomials: Theory and Practice. Kluwer Academic Publishers, Dordrecht, pp. 1–24, (1990).
2. M. Abramowitz and I.A. Stegun, *Handbook of Mathematical Functions with Formulas, Graphs, and Mathematical Tables, 9th printing*, New York: Dover, pp. 890 and 923, 1972.
3. F.V. Atkinson and W.N. Everitt, *Orthogonal polynomials which satisfy second order differential equations*, E. B. Christoffel (Aachen/Monschau, 1979), pp. 173–181, Birkhäuser, Basel-Boston, Mass., (1981).
4. W.C. Bauldry, *Estimates of asymmetric Freud polynomials on the real line*, J. Approx. Theory, 63(1990), 225–237.
5. W.C. Bauldry, *Orthogonal Polynomials Associated With Exponential Weights (Christoffel Functions, Recurrence Relations)*, Ph.D. thesis, The Ohio State University, ProQuest LLC, 149 pp., 1985.
6. A. Beiser, *Modern Physics*, 6 ed., McGraw-Hill, New York, 2003.
7. S.S. Bonan and D.S. Clark, *Estimates of the Hermite and the Freud polynomials*, J. Approx. Theory, 63(1990), 210–224.
8. S. Chandrasekhar, *Radiative Transfer*, Dover, New York, pp. 61 and 64–65, 1960.
9. Y. Chen and M.E.H Ismail, *Ladder operators and differential equations for orthogonal polynomials*, J. Phys. A, 30(1997), 7817–7829.
10. F.B. Hildebrand, *Introduction to Numerical Analysis*, McGraw-Hill, New York, pp. 338–343, 1956.
11. R. Koekoek and R.F. Swarttouw, *The Askey-scheme of hypergeometric orthogonal polynomials and its q-analogue*, Reports of the Faculty of Technical Mathematics and Information, No. 98–17, Delft University of Technology, (1998). http://aw.twi.tudelft.nl/~koekoek/askey/index.html
12. M.E.H. Ismail, *Classical and Quantum Orthogonal Polynomials in One Variable*, Encyclopedia of Mathematics and its Applications 98, Cambridge University Press, Cambridge, 2005.
13. M.E.H. Ismail, I. Nikolova and P. Simeonov, *Difference equations and discriminants for discrete orthogonal polynomials*, Ramanujan J., 8(2004), 475–502.
14. M.E.H. Ismail and J. Wimp, *On differential equations for orthogonal polynomials*, Methods Appl. Anal., 5(1998), 439–452.
15. J. Meixner, *Orthogonale Polynomsysteme mit einer besonderen Gestalt der erzeugenden Funktion*, J. London Math. Soc., 9(1934), 6–13.
16. H.N. Mhaskar, *Bounds for certain Freud-type orthogonal polynomials*, J. Approx. Theory, 63(1990), 238–254.
17. E.D. Rainville, *Special Functions*, Macmillan, New York, 1960.
18. E. Schrödinger, *An undulatory theory of the mechanics of atoms and molecules*, Phys. Rev., 28(1926), 1049–1070.
19. E. Schrödinger, *Quantisierung als eigenwertproblem*, Annalen der Physik, (1926), 361–377.
20. R. Shankar, *Principles of Quantum Mechanics*, 2nd ed., Plenum Publishers, 1994.
21. R.-C. Sheen, *Plancherel-Rotach-type asymptotics for orthogonal polynomials associated with* $\exp(-x^6/6)$, J. Approx. Theory, 50(1987), 232–293.
22. I.M. Sheffer, *Some properties of polynomial sets of type zero*, Duke Math J., 5(1939), 590–622.
23. J. Shohat, *A differential equation for orthogonal polynomials*, Duke Math. J., 5(1939), 401–417.
24. G. Szegő, *Orthogonal Polynomials*, fourth ed., American Mathematical Society, Colloquium Publications, Vol. XXIII, Providence, 1975.
25. E.T. Whittaker and G.N Watson, *Modern Analysis*, 4th ed., Cambridge, 1927.

Chapter 3
A Method for Analyzing a Special Case of the Sheffer B-Type 1 Polynomial Sequences

In 1939, I.M. Sheffer proved that every polynomial sequence belongs to one and only one *Type*. He also extensively developed properties of the most basic *Type* set, entitled *B-Type 0* or equivalently *A-Type 0* ($k = 0$ below)

$$\Lambda(t)\exp\left[xH_1(t) + \cdots + x^{k+1}H_{k+1}(t)\right] = \sum_{n=0}^{\infty} P_n(x)t^n,$$

with $H_i(t) = h_{i,i}t^i + h_{i,i+1}t^{i+1} + \cdots, \quad h_{1,1} \neq 0, \quad i = 1, 2, \ldots, k+1$

and then determined which of these sets are also orthogonal. He subsequently generalized his classification by letting k be arbitrary in the relation above, i.e., he defined the *B-Type k* class.

Thus far, no research has been published that *specifically* analyzes the higher-order Sheffer classes ($k \geq 1$ above). Therefore, we present a preliminary analysis of a special case of the *B-Type 1* ($k = 1$) class, in order to determine which sets, if any, are also orthogonal. Moreover, the method utilized herein is quite useful, as it can be applied to other types of characterization problems as well. This chapter also demonstrates how computer algebra packages, like Mathematica®, can aid in the development of rigorous results in orthogonal polynomials and special functions. We conclude this chapter by discussing some future research problems that can be solved by using the techniques of this chapter.

3.1 Preliminaries

Throughout this chapter, we make use of each of the following definitions, terminologies and notations.

Definition 3.1. We always assume that a *set* of polynomials $\{P_n(x)\}_{n=0}^{\infty}$ is such that each $P_n(x)$ has degree exactly n, which we write as $\deg(P_n(x)) = n$.

D.J. Galiffa, *On the Higher-Order Sheffer Orthogonal Polynomial Sequences*,   67
SpringerBriefs in Mathematics, DOI 10.1007/978-1-4614-5969-9_3,
© Daniel J. Galiffa 2013

Definition 3.2. The set of polynomials $\{P_n(x)\}_{n=0}^{\infty}$ is **orthogonal** if it satisfies one of the two weighted inner products below:

$$\text{Continuous}: \quad \langle P_m(x), P_n(x) \rangle = \int_{\Omega_1} P_m(x)P_n(x)w(x)\mathrm{d}x = \alpha_n \delta_{m,n},$$

$$\text{Discrete}: \quad \langle P_m(x), P_n(x) \rangle = \sum_{\Omega_2} P_m(x)P_n(x)w(x) = \beta_n \delta_{m,n},$$

where $\delta_{m,n}$ denotes the Kronecker delta

$$\delta_{m,n} := \begin{cases} 1 & \text{if} \quad m = n \\ 0 & \text{if} \quad m \neq n, \end{cases}$$

with $\Omega_1 \subseteq \mathbb{R}$, $\Omega_2 \subseteq \{0,1,2,\ldots\}$ and $w(x) > 0$ is entitled the **weight function**.

Definition 3.3. It is a necessary and sufficient condition that an orthogonal polynomial sequence $\{P_n(x)\}_{n=0}^{\infty}$ satisfies an **unrestricted three-term recurrence relation** of the form

$$P_{n+1}(x) = (A_n x + B_n)P_n(x) - C_n P_{n-1}(x), \quad A_n A_{n-1} C_n > 0$$

where $P_{-1}(x) = 0$ and $P_0(x) = 1$. \hfill (3.1)

We entitle $A_n A_{n-1} C_n > 0$ the **positivity condition**.

Definition 3.4. We shall define a **generating function** for a polynomial sequence $\{P_n(x)\}_{n=0}^{\infty}$ as follows:

$$\sum_{\Lambda} \zeta_n P_n(x)t^n = F(x,t),$$

with $\Lambda \subseteq \{0,1,2,\ldots\}$ and $\{\zeta_n\}_{n=0}^{\infty}$ a sequence in n that is independent of x and t. Moreover, we say that the function $F(x,t)$ **generates** the set $\{P_n(x)\}_{n=0}^{\infty}$.

3.2   The Formulation of Our Main Problem

In 1939, I.M. Sheffer published his seminal work [4], in which he basically showed that every polynomial sequence can be classified as belonging to exactly one *Type*. The majority of this paper was dedicated to developing a wealth of aesthetic results and interesting characterizing theorems regarding the most basic *Type* set, entitled *B-Type 0* or equivalently *A-Type 0*. Sheffer proved that a necessary and sufficient condition for a set $\{S_n(x)\}_{n=0}^{\infty}$ to be of *B-Type 0* is that it satisfies the generating function:

$$A(t)e^{xH(t)} = \sum_{n=0}^{\infty} S_n(x)t^n, \hfill (3.2)$$

where $A(t)$ and $H(t)$ are formal power series in t with $A(0) = 1$, $H(0) = 0$, and $H'(0) = 1$.

Moreover, one of Sheffer's most important results was his classification of the B-Type 0 orthogonal sets, which are often simply called the **Sheffer Sequences**. These sets are now known to be the very well-studied and applicable Laguerre, Hermite, Charlier, Meixner, Meixner–Pollaczek and Krawtchouk polynomials—refer to [3] for details regarding these polynomials and the references therein for additional theory and applications. Sheffer also subsequently generalized his classification to the case of arbitrary B-Type k by constructing the generalized generating function

$$A(t)\exp\left[xH_1(t) + \cdots + x^{k+1}H_{k+1}(t)\right] = \sum_{n=0}^{\infty} P_n(x)t^n,$$

with $H_i(t) = h_{i,i}t^i + h_{i,i+1}t^{i+1} + \cdots$, $h_{1,1} \neq 0$, $i = 1, 2, \ldots, k+1$. (3.3)

It is clear that (3.2) is a special case of (3.3).

In this chapter, we present an analysis of the B-Type 1 polynomials, $k = 1$ in (3.3):

$$A(t)\exp\left[xH_1(t) + x^2 H_2(t)\right] = \sum_{n=0}^{\infty} P_n(x)t^n,$$

with $H_i(t) = h_{i,i}t^i + h_{i,i+1}t^{i+1} + \cdots$, $h_{1,1} \neq 0$, $i = 1, 2$. (3.4)

in order to determine which, if any, of these sets are also orthogonal. This analysis is important for three primary reasons. One, the method that we utilize to analyze this problem can be applied to various *other* types of characterization problems. Two, for our method to be successful, computer algebra packages, like Mathematica®, are imperative. Therefore, we also demonstrate how such tools can be used to develop rigorous results in orthogonal polynomials and special functions. Three, no analysis has been published to date that *specifically* addresses the higher-order Sheffer classes.

To begin, since (3.1) is a necessary and sufficient condition for orthogonality, determining all sets $\{P_n(x)\}_{n=0}^{\infty}$ that arise from (3.4) and are also orthogonal is of course tantamount to finding all polynomial sequences that stem from (3.4) and also satisfy (3.1). Thus, in this section we develop a method for determining if conditions exist that guarantee that $\{P_n(x)\}_{n=0}^{\infty}$ simultaneously satisfies (3.4) and (3.1)—additional assumptions are made in Sect. 3.6. We now discuss the main elements of each of the sections that follow.

In Sect. 3.3, we determine the Sheffer B-Type 1 recursion coefficients A_n, B_n and C_n in (3.1) by analyzing (3.4). Section 3.4 amounts to determining the lower-order Sheffer B-Type 1 polynomials that arise from the generating function (3.4) and in Sect. 3.5, we establish the same polynomials from the three-term recurrence relationship (3.1), which have coefficients of a different structure than the ones achieved from the generating function (3.4). In Sect. 3.6, we first address the ramifications of the complexity of the B-Type 1 class. Then, we also show

how obtaining the aforementioned polynomials from Sects. 3.4 and 3.5 enables comparisons between certain terms to be made, yielding a simultaneous system of nonlinear algebraic equations, the solution(s) of which lead to conditions that must be satisfied in order for the sequence as defined by (3.4) to be orthogonal. In Sect. 3.7, we analyze each solution of the system obtained in Sect. 3.6 and 3.8 demonstrates how computer algebra can also be utilized to gain insights to the solution(s) of the system developed in Sect. 3.6. Lastly, in Sect. 3.9 we discuss some conclusions and future considerations.

It is also important to briefly discuss the role that symbolic computer algebra packages, like Mathematica® (refer to [5, 6]), play in this chapter and in the study of orthogonal polynomials in general. In fact, much of the computational aspects of this chapter were completed with Mathematica® and we emphasize that similar packages can of course also be used to complete the same computations. In addition, many contemporary mathematicians have used such packages to conduct research on orthogonal polynomials and special functions. For example, in [2] Gasper demonstrates how to use symbolic computer algebra packages to derive formulas involving orthogonal polynomials and other special functions; however, he also alludes to the fact that these packages can be used in many other areas of orthogonal polynomials as well. He says,

> Now that several symbolic computer algebra systems such as [Mathematica®] are available for various computers, it is natural for persons having access to such a system to try to have it perform the tedious manipulations needed to derive certain formulas involving orthogonal polynomials and special functions.

In our present work, the utilization of computer algebra packages, such as Mathematica®, is not simply a convenience but essentially a necessity, as the complexities involved in the computational aspects of this chapter are quite involved.

Herein, we clearly indicate when and how Mathematica® was used. For emphasis, we display the Mathematica® calculations and their respective outputs with a distinctive font class that is analogous to the style used in the Mathematica® notebook. In addition, we use the same conventions used in [2] for displaying a given Mathematica® input and its respective output. We demonstrate this notion with an example from Gasper's work.

Example 3.1. Gasper demonstrates a method for deriving a certain Laguerre polynomial expansion formula using Mathematica®. He displays his input and respective output in a similar manner as seen below:

$$In[5] : = p[b+1,n]p[-n,j]x^j/(p[1,n]p[1,j]p[b+1,j]),$$

$$Out[5] : = \frac{x^j p[-n,j]p[1+b,n]}{p[1,j]p[1,n]p[1+b,j]}.$$

Of course, the point of emphasis in the above example is not the actual computation but the fashion in which it is presented, as this type of display is consistent for all of the Mathematica® computations in this work.

3.3 The B-Type 1 Recursion Coefficients

We begin by constructing a method of effectively expanding the generating function (3.4), in order to acquire the coefficients of $x^n t^n$, $x^{n-1} t^n$ and $x^{n-2} t^n$. We do this since only these coefficients are needed to establish the recursion coefficients A_n, B_n and C_n in (3.1), which will become clear by the end of this section. First, we define $H_1(t) := H(t)$ and $H_2(t) := G(t)$ in (3.4) for ease of notation. Therefore, the Sheffer B-Type 1 polynomial sequences are defined as all of the polynomials sequences $\{P_n(x)\}_{n=0}^{\infty}$ that satisfy

$$A(t)\exp\left[xH(t) + x^2 G(t)\right] = \sum_{n=0}^{\infty} P_n(x)t^n. \tag{3.5}$$

Clearly, the left-hand side of (3.5) involves a power series expansion of power series expansions and can be written as

$$\sum_{i=0}^{\infty} a_i t^i \sum_{j=0}^{\infty} \frac{1}{j!} \left[x \sum_{k=1}^{\infty} h_k t^k + x^2 \sum_{l=2}^{\infty} g_l t^l \right]^j = \sum_{n=0}^{\infty} P_n(x)t^n.$$

This is certainly not the most advantageous way to perceive (3.5), but it does demonstrate its complexity. We ultimately expand (3.5) in a much more practical way. However, we first note that the initial assumptions of (3.2) require $a_0 = 1$ and $h_1 = 1$ and therefore, $A(t)$ and $H(t)$ respectively have the following structures:

$$A(t) = 1 + a_1 t + a_2 t^2 + \cdots \quad \text{and} \quad H(t) = t + h_2 t^2 + h_3 t^3 + \cdots.$$

Now, we expand the left-hand side of (3.5) in the following practical manner:

$$(1 + a_1 t + a_2 t^2 + \cdots) \left[e^{xt} e^{h_2 x t^2} e^{h_3 x t^3} \cdots \right] \left[e^{g_2 x^2 t^2} e^{g_3 x^2 t^3} e^{g_4 x^2 t^4} \cdots \right]. \tag{3.6}$$

Thus, expanding (3.6) in terms of the Maclaurin series of each product yields

$$\sum_{m=0}^{\infty} a_m t^m \prod_{i=1}^{\infty} \left[\sum_{j=0}^{\infty} \frac{h_i^j x^j t^{ij}}{j!} \right] \prod_{k=2}^{\infty} \left[\sum_{l=0}^{\infty} \frac{g_k^l x^{2l} t^{kl}}{l!} \right]. \tag{3.7}$$

With this convention, we write out the general term in each of the products of (3.7) as

$$a_{k_0} t^{k_0} \frac{x^{k_1} t^{k_1}}{k_1!} \frac{h_2^{k_2} x^{k_2} t^{2k_2}}{k_2!} \frac{h_3^{k_3} x^{k_3} t^{3k_3}}{k_3!} \cdots$$

$$\frac{g_2^{k_4} x^{2k_4} t^{2k_4}}{k_4!} \frac{g_3^{k_5} x^{2k_5} t^{3k_5}}{k_5!} \frac{g_4^{k_6} x^{2k_6} t^{4k_6}}{k_6!} \cdots, \tag{3.8}$$

where $\{k_0, k_1, k_2, \ldots\}$ are all nonnegative integers. It is important to note that in the expansion (3.8) we have only explicitly written the seven terms above since only these terms are needed to find the coefficients of $x^n t^n$, $x^{n-1} t^n$ and $x^{n-2} t^n$, as we explain below.

Notice, that the sums of the x-exponents and the t-exponents of (3.8), respectively, take on the form of the following linear Diophantine equations:

$$k_1 + k_2 + k_3 + 2k_4 + 2k_5 + 2k_6 + \cdots = r \tag{3.9}$$

and

$$k_0 + k_1 + 2k_2 + 3k_3 + 2k_4 + 3k_5 + 4k_6 + \cdots = s. \tag{3.10}$$

Explicitly writing out any additional terms in (3.9) or (3.10) is superfluous (as will be evident by the completion of Case 3 below) since upon subtracting (3.9) from (3.10) the additional terms will not contribute to finding the coefficients of $x^n t^n$, $x^{n-1} t^n$ and $x^{n-2} t^n$ because they will each be multiplied by a number greater than 2. As a consequence, it should be noted that in the computations that follow, $k_i = 0 \ \forall i \geq 7$. In addition, by subtracting (3.9) from (3.10), we immediately see that each of the coefficients in the polynomial $P_n(x)$ defined by (3.5) is a *finite sum*. This logic will be the basis of the analysis involved in obtaining the aforementioned coefficients $x^n t^n$, $x^{n-1} t^n$ and $x^{n-2} t^n$, which we partition into three parts.

(1) The Coefficient of $x^n t^n$

Upon subtracting (3.9) from (3.10) with $r = s = n$ we see that every k-value must be zero, except k_1 and k_4, which can take on any nonnegative integer values, i.e., are free variables. Therefore, by studying (3.8) we realize that the only terms involved in the coefficient of $x^n t^n$ are $1/k_1!$ and $g_2^{k_4}/k_4!$. Hence, by combining these two pieces of information, we observe that the coefficient of $x^n t^n$ must be a sum taken over all nonnegative integers k_1 and k_4 such that $k_1 + 2k_4 = n$ with argument $g_2^{k_4}/(k_1! k_4!)$, as seen below:

$$\sum_{k_1 + 2k_4 = n} \frac{g_2^{k_4}}{k_1! k_4!} \neq 0, \tag{3.11}$$

which is required to be nonzero via Definition 3.1.

(2) The Coefficient of $x^{n-1}t^n$

After subtracting (3.9) from (3.10) with $r = n - 1$ and $s = n$ we obtain

$$k_0 + k_2 + 2k_3 + k_5 + 2k_6 = 1. \tag{3.12}$$

Therefore, we see that $k_3 = k_6 = 0$ with k_1 and k_4 free variables, so (3.12) becomes

$$k_0 + k_2 + k_5 = 1,$$

yielding three cases.

Case 1: $k_0 = 1$, $k_2 = 0$, and $k_5 = 0$
In this case, we see that (3.10) turns into

$$1 + k_1 + 2k_4 = n,$$

resulting in

$$\sum_{k_1 + 2k_4 = n - 1} \frac{a_1 g_2^{k_4}}{k_1! k_4!}.$$

Case 2: $k_2 = 1$, $k_0 = 0$, and $k_5 = 0$
Here we see that (3.10) is now

$$k_1 + 2 + 2k_4 = n,$$

yielding

$$\sum_{k_1 + 2k_4 = n - 2} \frac{h_2 g_2^{k_4}}{k_1! k_4!}.$$

Case 3: $k_5 = 1$, $k_0 = 0$, and $k_2 = 0$
Lastly, we see that for this case (3.10) becomes

$$k_1 + 2k_4 + 3 = n,$$

leading to

$$\sum_{k_1 + 2k_4 = n - 3} \frac{g_3 g_2^{k_4}}{k_1! k_4!}$$

and we now have exhausted all possibilities.

Therefore, the coefficient of $x^{n-1}t^n$ is

$$\sum_{k_1+2k_4=n-1} \frac{a_1 g_2^{k_4}}{k_1! k_4!} + \sum_{k_1+2k_4=n-2} \frac{h_2 g_2^{k_4}}{k_1! k_4!} + \sum_{k_1+2k_4=n-3} \frac{g_3 g_2^{k_4}}{k_1! k_4!}. \tag{3.13}$$

(3) The Coefficient of $x^{n-2} t^n$

Here, after subtracting (3.9) from (3.10) with $r = n-2$ and $s = n$, we obtain

$$k_0 + k_2 + 2k_3 + k_5 + 2k_6 = 2.$$

Analogous to the previous coefficient derivations, the cases involved in finding the coefficient of $x^{n-2} t^n$ are determined by all of the nonnegative integer solutions to the equation directly above. In each case, the substitution of these solutions (the k-values) into (3.10) with $s = n$ is written as (i) and the resulting sum as (ii). We have a total of eight cases.

Case 1: $k_3 = 1$ and $k_0 = k_2 = k_5 = k_6 = 0$

$\qquad$ (i)$\quad$ $k_1 + 3 + 2k_4 = n$ $\qquad\qquad$ (ii)$\quad$ $\displaystyle\sum_{k_1+2k_4=n-3} \frac{h_3 g_2^{k_4}}{k_1! k_4!}$

Case 2: $k_6 = 1$ and $k_0 = k_2 = k_3 = k_5 = 0$

$\qquad$ (i)$\quad$ $k_1 + 2k_4 + 4 = n$ $\qquad\qquad$ (ii)$\quad$ $\displaystyle\sum_{k_1+2k_4=n-4} \frac{g_4 g_2^{k_4}}{k_1! k_4!}$

Case 3: $k_0 = 2$ and $k_2 = k_3 = k_5 = k_6 = 0$

$\qquad$ (i)$\quad$ $2 + k_1 + 2k_4 = n$ $\qquad\qquad$ (ii)$\quad$ $\displaystyle\sum_{k_1+2k_4=n-2} \frac{a_2 g_2^{k_4}}{k_1! k_4!}$

Case 4: $k_2 = 2$ and $k_0 = k_3 = k_5 = k_6 = 0$

$\qquad$ (i)$\quad$ $k_1 + 4 + 2k_4 = n$ $\qquad\qquad$ (ii)$\quad$ $\displaystyle\sum_{k_1+2k_4=n-4} \frac{h_2^2 g_2^{k_4}}{2! k_1! k_4!}$

Case 5: $k_5 = 2$ and $k_0 = k_2 = k_3 = k_6 = 0$

$\qquad$ (i)$\quad$ $k_1 + 2k_4 + 6 = n$ $\qquad\qquad$ (ii)$\quad$ $\displaystyle\sum_{k_1+2k_4=n-6} \frac{g_3^2 g_2^{k_4}}{2! k_1! k_4!}$

Case 6: $k_0 = k_2 = 1$ and $k_3 = k_5 = k_6 = 0$

$$(i) \quad 1 + k_1 + 2 + 2k_4 = n \qquad (ii) \quad \sum_{k_1 + 2k_4 = n-3} \frac{a_1 h_2 g_2^{k_4}}{k_1! k_4!}$$

Case 7: $k_0 = k_5 = 1$ and $k_2 = k_3 = k_6 = 0$

$$(i) \quad 1 + k_1 + 2k_4 + 3 = n \qquad (ii) \quad \sum_{k_1 + 2k_4 = n-4} \frac{a_1 g_3 g_2^{k_4}}{k_1! k_4!}$$

Case 8: $k_2 = k_5 = 1$ and $k_0 = k_3 = k_6 = 0$

$$(i) \quad k_1 + 2 + 2k_4 + 3 = n \qquad (ii) \quad \sum_{k_1 + 2k_4 = n-5} \frac{h_2 g_3 g_2^{k_4}}{k_1! k_4!}$$

Hence, the coefficient of $x^{n-2} t^n$ is

$$\sum_{k_1 + 2k_4 = n-3} \frac{h_3 g_2^{k_4}}{k_1! k_4!} + \sum_{k_1 + 2k_4 = n-4} \frac{g_4 g_2^{k_4}}{k_1! k_4!} + \sum_{k_1 + 2k_4 = n-2} \frac{a_2 g_2^{k_4}}{k_1! k_4!}$$

$$+ \sum_{k_1 + 2k_4 = n-4} \frac{h_2^2 g_2^{k_4}}{2! k_1! k_4!} + \sum_{k_1 + 2k_4 = n-6} \frac{g_3^2 g_2^{k_4}}{2! k_1! k_4!} + \sum_{k_1 + 2k_4 = n-3} \frac{a_1 h_2 g_2^{k_4}}{k_1! k_4!}$$

$$+ \sum_{k_1 + 2k_4 = n-4} \frac{a_1 g_3 g_2^{k_4}}{k_1! k_4!} + \sum_{k_1 + 2k_4 = n-5} \frac{h_2 g_3 g_2^{k_4}}{k_1! k_4!}. \tag{3.14}$$

We can then establish the following result.

Theorem 3.1. *The coefficients of x^n, x^{n-1} and x^{n-2} of $P_n(x)$ as defined by (3.5), respectively have the following form:*

$$c_{n,0} := \phi_n \neq 0, \tag{3.15}$$

$$c_{n,1} := a_1 \phi_{n-1} + h_2 \phi_{n-2} + g_3 \phi_{n-3}, \tag{3.16}$$

$$c_{n,2} := a_2 \phi_{n-2} + (h_3 + a_1 h_2) \phi_{n-3} + \left(g_4 + h_2^2/2! + a_1 g_3\right) \phi_{n-4}$$

$$+ h_2 g_3 \phi_{n-5} + (g_3^2/2!) \phi_{n-6}, \tag{3.17}$$

where

$$\phi_n(x) := \sum_{k=0}^{\lfloor n/2 \rfloor} \frac{x^k}{(n-2k)! k!}$$

with $\phi_n(g_2) := \phi_n$.

Proof. It can readily be shown that

$$\sum_{k_1+2k_4=n} \frac{g_2^{k_4}}{k_1!k_4!} = \sum_{k=0}^{\lfloor n/2 \rfloor} \frac{g_2^k}{(n-2k)!k!} = \phi_n.$$

Thus, upon manipulating the terms in (3.11), (3.13) and (3.14) accordingly we achieve our result.   □

We therefore conclude with the main result of this section.

Theorem 3.2. *The Sheffer B-Type 1 recursion coefficients A_n, B_n and C_n that satisfy* (3.1) *are given by*

$$A_n = \frac{c_{n+1,0}}{c_{n,0}}, \quad B_n = \frac{c_{n+1,1}c_{n,0} - c_{n+1,0}c_{n,1}}{c_{n,0}^2},$$

$$C_n = \frac{c_{n+1,0}(c_{n,0}c_{n,2} - c_{n,1}^2) + c_{n,0}(c_{n+1,1}c_{n,1} - c_{n+1,2}c_{n,0})}{c_{n-1,0}c_{n,0}^2}, \tag{3.18}$$

with $c_{n,0}$, $c_{n,1}$ and $c_{n,2}$, respectively, defined by (3.15), (3.16), *and* (3.17).

Proof. Based on Theorem 3.1, we see that $P_n(x)$ as defined by (3.5) now becomes

$$P_n(x) = c_{n,0}x^n + c_{n,1}x^{n-1} + c_{n,2}x^{n-2} + \mathcal{O}(x^{n-3})$$

and upon substituting $P_n(x)$ above into the three-term recurrence relation (3.1) we obtain

$$c_{n+1,0}x^{n+1} + c_{n+1,1}x^n + c_{n+1,2}x^{n-1} + \mathcal{O}(x^{n-2})$$
$$= A_n c_{n,0}x^{n+1} + A_n c_{n,1}x^n + A_n c_{n,2}x^{n-1} + \mathcal{O}(x^{n-2})$$
$$+ B_n c_{n,0}x^n + B_n c_{n,1}x^{n-1} + B_n c_{n,2}x^{n-2} + \mathcal{O}(x^{n-3})$$
$$- C_n c_{n-1,0}x^{n-1} - C_n c_{n-1,1}x^{n-2} - C_n c_{n-1,2}x^{n-3} + \mathcal{O}(x^{n-4}).$$

Thus, comparing the coefficients of x^{n+1}, x^n and x^{n-1} above results in the following lower-triangular simultaneous system of linear equations:

$$\begin{bmatrix} c_{n,0} & 0 & 0 \\ c_{n,1} & c_{n,0} & 0 \\ c_{n,2} & c_{n,1} & -c_{n-1,0} \end{bmatrix} \begin{bmatrix} A_n \\ B_n \\ C_n \end{bmatrix} = \begin{bmatrix} c_{n+1,0} \\ c_{n+1,1} \\ c_{n+1,2} \end{bmatrix}.$$

Since the diagonal terms $c_{n,0}$ and $c_{n-1,0}$ are nonzero, the solution to the above system is unique and determined to be (3.18) via elementary methods.   □

Remark 3.1. If we consider $\{Q_n(x)\}_{n=0}^{\infty}$ to be *any* orthogonal set satisfying (3.1), with leading coefficients, say $d_{n,0}$, $d_{n,1}$ and $d_{n,2}$, the corresponding recursion coefficients will of course have the same structure as those in Theorem 3.2. That is, the recursion coefficients will take on the forms below:

$$A_n = \frac{d_{n+1,0}}{d_{n,0}}, \quad B_n = \frac{d_{n+1,1}d_{n,0} - d_{n+1,0}d_{n,1}}{d_{n,0}^2}$$

$$C_n = \frac{d_{n+1,0}(d_{n,0}d_{n,2} - d_{n,1}^2) + d_{n,0}(d_{n+1,1}d_{n,1} - d_{n+1,2}d_{n,0})}{d_{n-1,0}d_{n,0}^2}.$$

3.4 Lower-Order B-Type 1 Polynomials Obtained via Generating Function

We can obtain any polynomial $P_n(x)$ by directly expanding (3.5). In order to determine $P_n(x)$, we first compute the coefficient of t^n on the left-hand side of (3.5) since this coefficient must be a polynomial in x and of degree n, which is clear since when writing (3.5) as

$$A(t)\exp\left[xH(t) + x^2 G(t)\right] = P_0(x) + P_1(x)t + P_2(x)t^2 + \cdots + P_n(x)t^n + \cdots$$

it is readily seen that the coefficient of t^n on the left-hand side of (3.5) must be $P_n(x)$. We now obtain the polynomials $P_0(x), \ldots, P_5(x)$.

The constant and linear polynomials are easily obtained by computations analogous to those previously described in Sect. 3.3 for finding the coefficients of $x^n t^n$, $x^{n-1}t^n$ and $x^{n-2}t^n$. For the constant polynomial $P_0(x)$, it is immediate that $a_0 = 1$ is the coefficient of t^0. Therefore, $P_0(x) = 1$. For the linear polynomial $P_1(x)$, we see again from direct computation that the coefficient of t^1 is $a_1 + a_0 h_1 x$ and thus $P_1(x) = a_1 + x$.

Of course, $P_2(x)$, $P_3(x)$ and the latter polynomials can be calculated in the same fashion. However, as would be the case for any polynomial sequence, the computations involved become increasingly more complicated as the degree increases, which, as we will show, is certainly evident in the Sheffer *B-Type 1* class. Therefore, in order to determine the polynomials $P_k(x)$ for $k \geq 2$ we have created a Mathematica® file entitled GenPoly with details addressed below.

We first define $h_1 := 1$. Then, we find the coefficient of t^2 by expanding the left-hand side of (3.5) via (3.7). Both the input and the respective output are seen below.

In[1]:= Expand[Coefficient[

$$\left(1 + \sum_{m=1}^{10} a_m t^m\right) * \prod_{j=1}^{10}\left(\sum_{i=0}^{10} \frac{h_j^i x^i t^{ij}}{i!}\right) * \prod_{k=2}^{10}\left(\sum_{l=0}^{10} \frac{g_k^l x^{2l} t^{kl}}{l!}\right), t^2\right]\right]$$

$$Out[1]= \frac{x^2}{2} + x a_1 + a_2 + x^2 g_2 + x h_2$$

Thus, we see that

$$P_2(x) = a_2 + (a_1 + h_2)x + \left(\frac{1}{2!} + g_2\right)x^2. \tag{3.19}$$

For the coefficient of t^3, we have the following computation and respective output:

$In[2]:=$ Expand[Coefficient[

$$\left(1 + \sum_{m=1}^{10} a_m t^m\right) * \prod_{j=1}^{10}\left(\sum_{i=0}^{10} \frac{h_j^i x^i t^{ji}}{i!}\right) * \prod_{k=2}^{10}\left(\sum_{l=0}^{10} \frac{g_k^l x^{2l} t^{kl}}{l!}\right), t^3\right]\right]$$

$$Out[2]= \frac{x^3}{6} + \frac{x^2 a_1}{2} + x a_2 + a_3 + x^3 g_2 + x^2 a_1 g_2 + x^2 g_3 + x^2 h_2 + x a_1 h_2 + x h_3$$

Therefore, we have

$$P_3(x) = a_3 + (a_2 + a_1 h_2 + h_3)x + \left(\frac{a_1}{2!} + a_1 g_2 + g_3 + h_2\right)x^2 + \left(\frac{1}{3!} + g_2\right)x^3. \tag{3.20}$$

It is important to mention that for the higher-order polynomials, specifically $P_6(x)$, $P_7(x)$ and $P_8(x)$, which will be computed in Sect. 3.6, the process is slightly adjusted since the outputs are more complicated than the ones above. Therefore, in order to more efficiently manage these polynomials, we first obtain the coefficient of t^n and then compute each x-coefficient individually. To demonstrate this procedure we construct $P_4(x)$.

We first find the coefficient of t^4:

$In[3]:=$ Expand[Coefficient[

$$\left(1 + \sum_{m=1}^{10} a_m t^m\right) * \prod_{j=1}^{10}\left(\sum_{i=0}^{10} \frac{h_j^i x^i t^{ji}}{i!}\right) * \prod_{k=2}^{10}\left(\sum_{l=0}^{10} \frac{g_k^l x^{2l} t^{kl}}{l!}\right), t^4\right]\right]$$

$$Out[3] = \frac{x^4}{24} + \frac{x^3 a_1}{6} + \frac{x^2 a_2}{2} + x a_3 + a_4 + \frac{x^4 g_2}{2} + x^3 a_1 g_2 + x^2 a_2 g_2 + \frac{x^4 g_2^2}{2} + x^3 g_3$$

$$+ x^2 a_1 g_3 + x^2 g_4 + \frac{x^3 h_2}{2} + x^2 a_1 h_2 + x a_2 h_2 + x^3 g_2 h_2 + \frac{x^2 h_2^2}{2} + x^2 h_3$$

$$+ x a_1 h_3 + x h_4.$$

For simplicity, we define the above output as follows:

$$In[4] := \text{FOURTH} := \frac{x^4}{24} + \frac{x^3 a_1}{6} + \frac{x^2 a_2}{2} + x a_3 + a_4 + \frac{x^4 g_2}{2} + x^3 a_1 g_2 + x^2 a_2 g_2$$

$$+ \frac{x^4 g_2^2}{2} + x^3 g_3 + x^2 a_1 g_3 + x^2 g_4 + \frac{x^3 h_2}{2} + x^2 a_1 h_2 + x a_2 h_2 + x^3 g_2 h_2$$

$$+ \frac{x^2 h_2^2}{2} + x^2 h_3 + x a_1 h_3 + x h_4.$$

Then we compute each coefficient separately.
    For the coefficient of x^4 we have

$$In[5] := \text{Coefficient}[\text{FOURTH, } x^4]$$

$$Out[5] = \frac{1}{24} + \frac{g_2}{2} + \frac{g_2^2}{2}$$

Then, for the coefficient of x^3, we see that

$$In[6] := \text{Coefficient}[\text{FOURTH, } x^3]$$

$$Out[6] = \frac{a_1}{6} + a_1 g_2 + g_3 + \frac{h_2}{2} + g_2 h_2$$

Next, the coefficient of x^2 is computed as

$$In[7] := \text{Coefficient}[\text{FOURTH, } x^2]$$

$$Out[7] = \frac{a_2}{2} + a_2 g_2 + a_1 g_3 + g_4 + a_1 h_2 + \frac{h_2^2}{2} + h_3.$$

For the coefficient of x we obtain

$$In[8] := \text{Coefficient}[\text{FOURTH, } x]$$

$$Out[8] = a_3 + a_2 h_2 + a_1 h_3 + h_4$$

and for the constant term we achieve

$In[9]:= \texttt{Coefficient}[\texttt{x*FOURTH, x}]$

$$Out[9]= a_4.$$

Hence, putting these above pieces together we have

$$P_4(x) = a_4 + (a_3 + a_2 h_2 + a_1 h_3 + h_4)x$$

$$+ \left(\frac{a_2}{2!} + a_2 g_2 + a_1 g_3 + g_4 + a_1 h_2 + \frac{h_2^2}{2!} + h_3 \right) x^2$$

$$+ \left(\frac{a_1}{3!} + a_1 g_2 + g_3 + \frac{h_2}{2!} + g_2 h_2 \right) x^3 + \left(\frac{1}{4!} + \frac{g_2}{2!} + \frac{g_2^2}{2!} \right) x^4. \qquad (3.21)$$

Using the same process as demonstrated above, we also obtain an expression for $P_5(x)$.

$$P_5(x) = a_5 + (a_4 + a_3 h_2 + a_2 h_3 + a_1 h_4 + h_5)x$$

$$+ \left(\frac{a_3}{2!} + a_3 g_2 + a_2 g_3 + a_1 g_4 + g_5 + a_2 h_2 + \frac{a_1 h_2^2}{2!} + a_1 h_3 + h_2 h_3 + h_4 \right) x^2$$

$$+ \left(\frac{a_2}{3!} + a_2 g_2 + a_1 g_3 + g_4 + \frac{a_1 h_2}{2!} + a_1 g_2 h_2 + g_3 h_2 + \frac{h_2^2}{2!} + \frac{h_3}{2!} + g_2 h_3 \right) x^3$$

$$+ \left(\frac{a_1}{4!} + \frac{a_1 g_2}{2!} + \frac{a_1 g_2^2}{2!} + \frac{g_3}{2!} + g_2 g_3 + \frac{h_2}{3!} + g_2 h_2 \right) x^4 + \left(\frac{1}{5!} + \frac{g_2}{3!} + \frac{g_2^2}{2!} \right) x^5.$$

3.5 Lower-Order B-Type 1 Polynomials Obtained via Three-Term Recurrence Relation

We next find the polynomials $P_0(x),\ldots,P_5(x)$ from the three-term recurrence relation (3.1) with A_n, B_n and C_n as defined in (3.18). To accomplish this, we have created a Mathematica® file entitled $\texttt{ThreeTerm}$, the results of which are utilized throughout this section. We first define $c_{n,0}$, $c_{n,1}$ and $c_{n,2}$, as, respectively, established in (3.15), (3.16) and (3.17):

$$In\,[1]:= c_0[n\_] := \sum_{k=0}^{\text{Floor}[n/2]} \frac{g_2^k}{(n-2k)!\,k!}$$

$$In[2]:= c_1[n\_] := \sum_{k=0}^{\text{Floor}[(n-1)/2]} \frac{a_1 * g_2^k}{(n-1-2k)!k!} + \sum_{k=0}^{\text{Floor}[(n-2)/2]} \frac{h_2 * g_2^k}{(n-2-2k)!k!}$$

$$+ \sum_{k=0}^{\text{Floor}[(n-3)/2]} \frac{g_3 * g_2^k}{(n-3-2k)!k!}$$

$$In[3]:= c_2[n\_] := \sum_{k=0}^{\text{Floor}[(n-2)/2]} \frac{a_2 * g_2^k}{(n-2-2k)!k!}$$

$$+ (h_3 + a_1 * h_2) * \sum_{k=0}^{\text{Floor}[(n-3)/2]} \frac{g_2^k}{(n-3-2k)!k!}$$

$$+ \left(g_4 + h_2^2/2! + a_1 * g_3\right) * \sum_{k=0}^{\text{Floor}[(n-4)/2]} \frac{g_2^k}{(n-4-2k)!k!}$$

$$+ \sum_{k=0}^{\text{Floor}[(n-5)/2]} \frac{h_2 * g_3 * g_2^k}{(n-5-2k)!k!} + \sum_{k=0}^{\text{Floor}[(n-6)/2]} \frac{g_3^2 * g_2^k}{2!(n-6-2k)!k!}$$

Then we define the A_n, B_n and C_n, as derived in (3.18).

$$In[4]:= A[n\_] := \frac{c_0[n+1]}{c_0[n]},$$

$$In[5]:= B[n\_] := \frac{c_1[n+1] * c_0[n] - c_0[n+1]c_1[n]}{c_0[n]^2},$$

$$In[6]:= C[n\_] := \frac{1}{c_0[n-1] * c_0[n]^2} \left(c_0[n+1] * \left(c_0[n] * c_2[n] - c_1[n]^2 \right) \right.$$

$$\left. + c_0[n] * (c_1[n+1] * c_1[n] - c_2[n+1] * c_0[n]) \right).$$

Lastly, in accordance with Sect. 3.4, we assign the constant and linear polynomials as seen below:

$$P_0 := 1 \text{ and } P_1 := a_1 + x.$$

Thus, we can now produce any polynomial defined by (3.5) of degree greater than one. As an example of the process, we achieve $P_2(x)$ by separately computing the quadratic, linear and constant terms and then amalgamate the results, as seen below:

$$In[7]:= \texttt{Coefficient}[x * A[1] * P_1 + B[1] * P_1 - C[1] * P_0, \ x^2]$$

$$Out[7]= \frac{1}{2!} + g_2.$$

$In[8]:=$ Together[Coefficient[$x * A[1] * P_1 + B[1] * P_1 - C[1] * P_0$, x]]

$$Out[8]= a_1 + h_2.$$

$In[9]:=$ Together[Coefficient[$x * (x * A[1] * P_1 + B[1] * P_1 - C[1] * P_0)$, x]]

$$Out[9]= a_2.$$

Therefore,

$$P_2(x) = a_2 + (a_1 + h_2)x + \left(\frac{1}{2!} + g_2 \right) x^2,$$

which is equal to (3.19).

Continuing in the same manner, we determine that the cubic, quadratic and linear terms of $P_3(x)$ as obtained in ThreeTerm coincide exactly with those in (3.20). However, the constant term is quite different, as seen below:

$In[10]:=$ Together[Coefficient[$x * (x * A[2] * P_2 + B[2] * P_2 - C[2] * P_1)$, x]]

$$Out[10]= \frac{1}{3(1+2g_2)^2} \left(-a_1^3 + 3a_1 a_2 + 4a_1 a_2 g_2 - 12a_1^3 g_2^2 + 12a_1 a_2 g_2^2 - 6a_1^2 g_3 \right.$$
$$+ 6a_2 g_3 - 12a_1^2 g_2 g_3 + 12a_2 g_2 g_3 - 2a_1^2 h_2 + 4a_2 h_2 + 12a_1^2 g_2 h_2 - 6a_1 g_3 h_2$$
$$\left. - 12a_1 g_2 g_3 h_2 - 4a_1 h_2^2 + 3a_1 h_3 + 12a_1 g_2 h_3 + 12a_1 g_2^2 h_3 \right).$$

Therefore, we have developed a relationship for the constant term a_3, since for $\{P_n(x)\}_{n=0}^{\infty}$ as defined by (3.5) to be orthogonal it must be that

$$a_3 = \frac{1}{3(1+2g_2)^2} \left(-a_1^3 + 3a_1 a_2 + 4a_1 a_2 g_2 - 12a_1^3 g_2^2 + 12a_1 a_2 g_2^2 - 6a_1^2 g_3 + 6a_2 g_3 \right.$$
$$- 12a_1^2 g_2 g_3 + 12a_2 g_2 g_3 - 2a_1^2 h_2 + 4a_2 h_2 + 12a_1^2 g_2 h_2 - 6a_1 g_3 h_2 - 12a_1 g_2 g_3 h_2$$
$$\left. - 4a_1 h_2^2 + 3a_1 h_3 + 12a_1 g_2 h_3 + 12a_1 g_2^2 h_3 \right). \tag{3.22}$$

For $P_4(x)$ we see that the fourth-degree term, the cubic term and the quadratic term are identical to those in (3.21); however, the linear and constant terms are dissimilar.

The linear term is as follows

$In[11]:=$ Together[Coefficient[$x * A[3] * P_3 + B[3] * P_3 - C[3] * P_2$, x]]

$Out[11]= \dfrac{1}{4\,(1+2g_2)\,(1+6g_2)^2}\,(-a_1^3 + 3a_1a_2 + a_3 - 8a_1^3g_2 + 28a_1a_2g_2 + 20a_3g_2$

$\qquad -48a_1^3g_2^2 + 120a_1a_2g_2^2 + 120a_3g_2^2 + 144a_1a_2g_2^3 + 240a_3g_2^3 + 144a_1^3g_2^4$

$\qquad -144a_1a_2g_2^4 + 144a_3g_2^4 - 2a_1^2h_2 + 8a_2h_2 + 12a_1^2g_2h_2 + 56a_2g_2h_2$

$\qquad -72a_1^2g_2^2h_2 + 192a_2g_2^2h_2 - 144a_1^2g_2^3h_2 + 288a_2g_2^3h_2 - 4a_1h_2^2$

$\qquad +48a_1g_2h_2^2 + 48a_1g_2^2h_2^2 - 8h_2^3 + 7a_1h_3 + 68a_1g_2h_3 + 168a_1g_2^2h_3$

$\qquad -144a_1g_2^3h_3 - 144a_1g_2^4h_3 + 12h_2h_3 + 96g_2h_2h_3 + 240g_2^2h_2h_3 + 48g_3h_2^2$

$\qquad -12a_1^2g_3 + 18a_2g_3 - 24a_1^2g_2g_3 + 108a_2g_2g_3 + 144a_1^2g_2^2g_3 + 72a_2g_2^2g_3$

$\qquad +288a_1^2g_2^3g_3 - 144a_2g_2^3g_3 - 36a_1g_3^2 - 144a_1g_2g_3^2 + 144a_1g_2^2g_3^2 + 8a_1g_4$

$\qquad +96a_1g_2g_4 + 288a_1g_2^2g_4 - 42a_1g_3h_2 - 108a_1g_2g_3h_2 + 216a_1g_2^2g_3h_2$

$\qquad +144a_1g_2^3g_3h_2 - 36g_3^2h_2 - 144g_2g_3^2h_2 + 144g_2^2g_3^2h_2 + 8g_4h_2 + 96g_2g_4h_2$

$\qquad +288g_2^2g_4h_2 - 192g_2g_3h_2^2 + 18g_3h_3 + 108g_2g_3h_3 + 72g_2^2g_3h_3 - 144g_2^3g_3h_3)$

and the $P_4(x)$ constant-term computation turns out to be:

$In[12]:=$ $\texttt{Together[Coefficient[x * (x * A[3] * P_3 + B[3] * P_3 - C[3] * P_2),}$
$x]]$

$Out[12]= \dfrac{1}{4\,(1+2g_2)\,(1+6g_2)^2}\,(-a_1^2a_2 + 2a_2^2 + a_1a_3 - 8a_1^2a_2g_2 + 20a_2^2g_2$

$\qquad +8a_1a_3g_2 - 48a_1^2a_2g_2^2 + 72a_2^2g_2^2 + 48a_1a_3g_2^2 + 144a_2^2g_2^3 + 144a_1^2a_2g_2^4$

$\qquad -144a_1a_3g_2^4 - 2a_1a_2h_2 + 6a_3h_2 + 12a_1a_2g_2h_2 + 36a_3g_2h_2 - 72a_1a_2g_2^2h_2$

$\qquad +120a_3g_2^2h_2 - 144a_1a_2g_2^3h_2 + 144a_3g_2^3h_2 - 8a_2h_2^2 + 6a_2h_3 + 60a_2g_2h_3$

$\qquad +120a_2g_2^2h_3 - 144a_2g_2^3h_3 - 12a_1a_2g_3 + 18a_3g_3 - 24a_1a_2g_2g_3$

$\qquad +108a_3g_2g_3 + 144a_1a_2g_2^2g_3 + 72a_3g_2^2g_3 + 288a_1a_2g_2^3g_3 - 144a_3g_2^3g_3$

$\qquad -36a_2g_3^2 - 144a_2g_2g_3^2 + 144a_2g_2^2g_3^2 + 8a_2g_4 + 96a_2g_2g_4$

$\qquad -48a_2g_3h_2 - 192a_2g_2g_3h_2 + 288a_2g_2^2g_4)$

Thus, appropriately equating the linear and constant terms above with the linear and constant terms of (3.21), we see that the linear term comparison is

$a_3 + a_2h_2 + a_1h_3 + h_4$

$$
\begin{aligned}
= \frac{1}{4\left(1+2g_2\right)\left(1+6g_2\right)^2} \bigl(&-a_1^3+3a_1a_2+a_3-8a_1^3g_2+28a_1a_2g_2+20a_3g_2 \\
&-48a_1^3g_2^2+120a_1a_2g_2^2+120a_3g_2^2+144a_1a_2g_2^3+240a_3g_2^3+144a_1^3g_2^4 \\
&-144a_1a_2g_2^4+144a_3g_2^4-2a_1^2h_2+8a_2h_2+12a_1^2g_2h_2+56a_2g_2h_2 \\
&-72a_1^2g_2^2h_2+192a_2g_2^2h_2-144a_1^2g_2^3h_2+288a_2g_2^3h_2-4a_1h_2^2 \\
&+48a_1g_2h_2^2+48a_1g_2^2h_2^2-8h_2^3+7a_1h_3+68a_1g_2h_3+168a_1g_2^2h_3 \\
&-144a_1g_2^3h_3-144a_1g_2^4h_3+12h_2h_3+96g_2h_2h_3+240g_2^2h_2h_3+48g_3h_2^2 \\
&-12a_1^2g_3+18a_2g_3-24a_1^2g_2g_3+108a_2g_2g_3+144a_1^2g_2^2g_3+72a_2g_2^2g_3 \\
&+288a_1^2g_2^3g_3-144a_2g_2^3g_3-36a_1g_3^2-144a_1g_2g_3^2+144a_1g_2^2g_3^2+8a_1g_4 \\
&+96a_1g_2g_4+288a_1g_2^2g_4-42a_1g_3h_2-108a_1g_2g_3h_2+216a_1g_2^2g_3h_2 \\
&+144a_1g_2^3g_3h_2-36g_3^2h_2-144g_2g_3^2h_2+144g_2^2g_3^2h_2+8g_4h_2+96g_2g_4h_2 \\
&+288g_2^2g_4h_2-192g_2g_3h_2^2+18g_3h_3+108g_2g_3h_3+72g_2^2g_3h_3-144g_2^3g_3h_3\bigr)
\end{aligned}
$$

$$(3.23)$$

and the constant-term comparison is

$$
\begin{aligned}
a_4 = \frac{1}{4\left(1+2g_2\right)\left(1+6g_2\right)^2} \bigl(&-a_1^2a_2+2a_2^2+a_1a_3-8a_1^2a_2g_2+20a_2^2g_2 \\
&+8a_1a_3g_2-48a_1^2a_2g_2^2+72a_2^2g_2^2+48a_1a_3g_2^2+144a_2^2g_2^3+144a_1^2a_2g_2^4 \\
&-144a_1a_3g_2^4-2a_1a_2h_2+6a_3h_2+12a_1a_2g_2h_2+36a_3g_2h_2-72a_1a_2g_2^2h_2 \\
&+120a_3g_2^2h_2-144a_1a_2g_2^3h_2+144a_3g_2^3h_2-8a_2h_2^2+6a_2h_3+60a_2g_2h_3 \\
&+120a_2g_2^2h_3-144a_2g_2^3h_3-12a_1a_2g_3+18a_3g_3-24a_1a_2g_2g_3+108a_3g_2g_3 \\
&+144a_1a_2g_2^2g_3+72a_3g_2^2g_3+288a_1a_2g_2^3g_3-144a_3g_2^3g_3-36a_2g_3^2-144a_2g_2g_3^2 \\
&+144a_2g_2^2g_3^2+8a_2g_4+96a_2g_2g_4-48a_2g_3h_2-192a_2g_2g_3h_2+288a_2g_2^2g_4\bigr).
\end{aligned}
$$

$$(3.24)$$

Now, continuing this process for $P_5(x)$, we obtain three additional relationships for the $P_5(x)$ quadratic, linear and constant terms. These relationships are displayed below. First, we have the $P_5(x)$ quadratic-term comparison:

$$
\begin{aligned}
\frac{a_3}{2!} \; +a_3g_2 & +a_2g_3+a_1g_4+g_5+a_2g_2+\frac{a_1h_2^2}{2!}+a_1h_3+h_2h_3+h_4 \\
& = \bigl(-a_1^3+3a_1a_2+2a_3-24a_1^3g_2+76a_1a_2g_2+76a_3g_2-260a_1^3g_2^2
\end{aligned}
$$

$$+788a_1a_2g_2^2 + 1008a_3g_2^2 - 1152a_1^3g_2^3 + 3744a_1a_2g_2^3 + 5664a_3g_2^3 - 2160a_1^3g_2^4$$

$$+7056a_1a_2g_2^4 + 12960a_3g_2^4 - 5760a_1^3g_2^5 + 8640a_1a_2g_2^5 + 8640a_3g_2^5 - 8640a_1^3g_2^6$$

$$+8640a_1a_2g_2^6 - 2a_1^2h_2 + 14a_2h_2 - 4a_1^2g_2h_2 + 332a_2g_2h_2 - 144a_1^2g_2^2h_2$$

$$+3216a_2g_2^2h_2 - 288a_1^2g_2^3h_2 + 14112a_2g_2^3h_2 + 4320a_1^2g_2^4h_2 + 21600a_2g_2^4h_2$$

$$+8640a_1^2g_2^5h_2 + 8640a_2g_2^5h_2 + a_1h_2^2 + 214a_1g_2h_2^2 + 1464a_1g_2^2h_2^2 + 4176a_1g_2^3h_2^2$$

$$+6480a_1g_2^4h_2^2 + 4320a_1g_2^5h_2^2 - 16h_2^3 + 192g_2^2h_2^3 + 13a_1h_3 + 336a_1g_2h_3 + 3204a_1g_2^2h_3$$

$$+13440a_1g_2^3h_3 + 28080a_1g_2^4h_3 + 34560a_1g_2^5h_3 + 8640a_1g_2^6h_3 + 34h_2h_3 + 684g_2h_2h_3$$

$$+4848g_2^2h_2h_3 + 12960g_2^3h_2h_3 + 12960g_2^4h_2h_3 + 8640g_2^5h_2h_3 + 2h_4 + 76g_2h_4$$

$$+1008g_2^2h_4 + 5664g_2^3h_4 + 12960g_2^4h_4 + 8640g_2^5h_4 - 78a_1g_3h_2 + 2880g_2^3g_3h_2^2$$

$$-18a_1^2g_3 + 40a_2g_3 - 180a_1^2g_2g_3 + 744a_2g_2g_3 - 720a_1^2g_2^2g_3$$

$$+4896a_2g_2^2g_3 - 3744a_1^2g_2^3g_3 + 14976a_2g_2^3g_3 - 18720a_1^2g_2^4g_3 + 23040a_2g_2^4g_3$$

$$-25920a_1^2g_2^5g_3 + 17280a_2g_2^5g_3 - 180a_1g_3^2 - 1728a_1g_2g_3^2 - 3744a_1g_2^2g_3^2$$

$$-11520a_1g_2^3g_3^2 - 25920a_1g_2^4g_3^2 - 432g_3^3 - 4320g_2g_3^3 - 8640g_2^2g_3^3$$

$$-17280g_2^3g_3^3 + 26a_1g_4 + 620a_1g_2g_4 + 4656a_1g_2^2g_4 + 10656a_1g_2^3g_4$$

$$+7200a_1g_2^4g_4 + 8640a_1g_2^5g_4 + 120g_3g_4 + 2208g_2g_3g_4 + 12096g_2^2g_3g_4$$

$$+17280g_2^3g_3g_4 + 17280g_2^4g_3g_4 - 252a_1g_2g_3h_2 + 2256a_1g_2^2g_3h_2 + 7200a_1g_2^3g_3h_2$$

$$-4320a_1g_2^4g_3h_2 - 8640a_1g_2^5g_3h_2 - 684g_3^2h_2 - 6624g_2g_3^2h_2 - 12960g_2^2g_3^2h_2$$

$$-17280g_2^3g_3^2h_2 - 8640g_2^4g_3^2h_2 + 64g_4h_2 + 1344g_2g_4h_2 + 8448g_2^2g_4h_2$$

$$+11520g_2^3g_4h_2 - 240g_3h_2^2 - 1872g_2g_3h_2^2 - 2304g_2^2g_3h_2^2 + 90g_3h_3$$

$$+1548g_2g_3h_3 + 8496g_2^2g_3h_3 + 18720g_2^3g_3h_3 + 30240g_2^4g_3h_3 + 8640g_2^5g_3h_3 \Big)$$

$$/ \left(10(1+6g_2)\left(1+12g_2+12g_2^2\right)^2 \right).$$

(3.25)

Next, we have the $P_5(x)$ linear-term comparison

$$a_4 + a_3h_2 + a_2h_3 + a_1h_4 + h_5$$

$$= \Big(-a_1^2a_2 + 2a_2^2 + a_1a_3 + a_4 - 22a_1^2a_2g_2 + 48a_2^2g_2 + 22a_1a_3g_2 + 38a_4g_2$$

$$-216a_1^2a_2g_2^2 + 432a_2^2g_2^2 + 216a_1a_3g_2^2 + 504a_4g_2^2 - 720a_1^2a_2g_2^3$$

$$+1728a_2^2g_2^3 + 720a_1a_3g_2^3 + 2832a_4g_2^3 - 720a_1^2a_2g_2^4 + 1440a_2^2g_2^4$$

$$+720a_1a_3g_2^4 + 6480a_4g_2^4 - 4320a_1^2a_2g_2^5 + 4320a_1a_3g_2^5 + 4320a_4g_2^5$$

$$-a_1^3h_2 + a_1a_2h_2 + 8a_3h_2 - 22a_1^3g_2h_2 + 70a_1a_2g_2h_2$$

$$+144a_3g_2h_2 - 216a_1^3g_2^2h_2 + 504a_1a_2g_2^2h_2 + 1056a_3g_2^2h_2 - 720a_1^3g_2^3h_2$$

$$+2448a_1a_2g_2^3h_2 + 2880a_3g_2^3h_2 - 720a_1^3g_2^4h_2 + 6480a_1a_2g_2^4h_2 - 4320a_1^3g_2^5h_2$$

$$+4320a_1a_2g_2^5h_2 - 2a_1^2h_2^2 - 4a_2h_2^2 + 72a_2g_2h_2^2 - 144a_1^2g_2^2h_2^2 + 624a_2g_2^2h_2^2$$

$$+1440a_2g_2^3h_2^2 + 4320a_1^2g_2^4h_2^2 - 12a_1h_2^3 - 72a_1g_2h_2^3 - 432a_1g_2^2h_2^3$$

$$-1440a_1g_2^3h_2^3 + 11a_2h_3 + 246a_2g_2h_3 + 1800a_2g_2^2h_3 + 5328a_2g_2^3h_3$$

$$+7920a_2g_2^4h_3 + 4320a_2g_2^5h_3 + 15a_1h_2h_3 + 342a_1g_2h_2h_3 + 2280a_1g_2^2h_2h_3$$

$$+6480a_1g_2^3h_2h_3 + 10800a_1g_2^4h_2h_3 + 4320a_1g_2^5h_2h_3 - 12h_2^2h_3 - 72g_2h_2^2h_3$$

$$-432g_2^2h_2^2h_3 - 1440g_2^3h_2^2h_3 + 9h_3^2 + 198g_2h_3^2 + 1368g_2^2h_3^2$$

$$+3600g_2^3h_3^2 + 6480g_2^4h_3^2 + 4320g_2^5h_3^2 + a_1h_4 + 22a_1g_2h_4 - 126a_1g_3h_2^2$$

$$+216a_1g_2^2h_4 + 720a_1g_2^3h_4 + 720a_1g_2^4h_4 + 4320a_1g_2^5h_4 + 8h_2h_4$$

$$-18a_1a_2g_3 + 36a_3g_3 - 144a_1a_2g_2g_3 + 576a_3g_2g_3 - 432a_1a_2g_2^2g_3$$

$$+2880a_3g_2^2g_3 - 2880a_1a_2g_2^3g_3 + 5760a_3g_2^3g_3 - 12960a_1a_2g_2^4g_3 + 8640a_3g_2^4g_3$$

$$-216a_2g_3^2 - 2160a_2g_2g_3^2 - 4320a_2g_2^2g_3^2 - 8640a_2g_2^3g_3^2 + 24a_2g_4$$

$$+528a_2g_2g_4 + 3168a_2g_2^2g_4 + 2880a_2g_2^3g_4 - 18a_1^2g_3h_2 - 90a_2g_3h_2$$

$$-144a_1^2g_2g_3h_2 - 576a_2g_2g_3h_2 - 432a_1^2g_2^2g_3h_2 + 720a_2g_2^2g_3h_2 - 2880a_1^2g_2^3g_3h_2$$

$$+5760a_2g_2^3g_3h_2 - 4320g_2^4g_3h_2h_3 - 12960a_1^2g_2^4g_3h_2 + 4320a_2g_2^4g_3h_2 - 216a_1g_3^2h_2$$

$$-2160a_1g_2g_3^2h_2 - 4320a_1g_2^2g_3^2h_2 - 8640a_1g_2^3g_3^2h_2 + 24a_1g_4h_2 + 528a_1g_2g_4h_2$$

$$+3168a_1g_2^2g_4h_2 + 2880a_1g_2^3g_4h_2 - 1152a_1g_2g_3h_2^2 - 2160a_1g_2^2g_3h_2^2 - 4320a_1g_2^3g_3h_2^2$$

$$+18a_1g_3h_3 + 432a_1g_2g_3h_3 + 2448a_1g_2^2g_3h_3 + 2880a_1g_2^3g_3h_3 - 4320a_1g_2^4g_3h_3$$

$$-216g_3^2h_3 - 2160g_2g_3^2h_3 - 4320g_2^2g_3^2h_3 - 8640g_2^3g_3^2h_3 + 24g_4h_3$$

$$+528g_2g_4h_3 + 3168g_2^2g_4h_3 + 2880g_2^3g_4h_3 - 126g_3h_2h_3 - 1152g_2g_3h_2h_3$$

$$-2160g_2^2g_3h_2h_3 + 36g_3h_4 + 576g_2g_3h_4 + 2880g_2^2g_3h_4 + 5760g_2^3g_3h_4$$

$$+8640g_2^4g_3h_4 + 144g_2h_2h_4 + 1056g_2^2h_2h_4$$

$$+2880g_2^3h_2h_4 \Big) / \left(5\left(1 + 6g_2\right)\left(1 + 12g_2 + 12g_2^2\right)^2 \right) \tag{3.26}$$

and finally, the $P_5(x)$ constant-term comparison is

$$
\begin{aligned}
a_5 =\bigl(&-a_1^2a_3 + 2a_2a_3 + a_1a_4 - 22a_1^2a_3g_2 + 48a_2a_3g_2 \\
&+22a_1a_4g_2 - 216a_1^2a_3g_2^2 + 432a_2a_3g_2^2 + 216a_1a_4g_2^2 - 720a_1^2a_3g_2^3 \\
&+1728a_2a_3g_2^3 + 720a_1a_4g_2^3 - 720a_1^2a_3g_2^4 + 1440a_2a_3g_2^4 + 720a_1a_4g_2^4 \\
&-4320a_1^2a_3g_2^5 + 4320a_1a_4g_2^5 - 2a_1a_3h_2 + 8a_4h_2 + 144a_4g_2h_2 \\
&-144a_1a_3g_2^2h_2 + 1056a_4g_2^2h_2 + 2880a_4g_2^3h_2 + 4320a_1a_3g_2^4h_2 - 12a_3h_2^2 \\
&-72a_3g_2h_2^2 - 432a_3g_2^2h_2^2 - 1440a_3g_2^3h_2^2 + 9a_3h_3 + 198a_3g_2h_3 \\
&+1368a_3g_2^2h_3 + 3600a_3g_2^3h_3 + 6480a_3g_2^4h_3 + 4320a_3g_2^5h_3 \\
&-18a_1a_3g_3 + 36a_4g_3 - 144a_1a_3g_2g_3 + 576a_4g_2g_3 - 432a_1a_3g_2^2g_3 \\
&+2880a_4g_2^2g_3 - 2880a_1a_3g_2^3g_3 + 5760a_4g_2^3g_3 - 12960a_1a_3g_2^4g_3 \\
&+8640a_4g_2^4g_3 - 216a_3g_3^2 - 2160a_3g_2g_3^2 - 4320a_3g_2^2g_3^2 - 8640a_3g_2^3g_3^2 \\
&+24a_3g_4 + 528a_3g_2g_4 + 3168a_3g_2^2g_4 + 2880a_3g_2^3g_4 - 126a_3g_3h_2 \\
&-1152a_3g_2g_3h_2 - 2160a_3g_2^2g_3h_2 - 4320a_3g_2^4g_3h_2 \\
&+4320a_3g_2^5h_3\bigr) \Big/ \Bigl(5\left(1 + 6g_2\right)\left(1 + 12g_2 + 12g_2^2\right)^2\Bigr).
\end{aligned}
\tag{3.27}
$$

Notice that in each of the coefficient comparisons above, the x^n, x^{n-1} and x^{n-2} comparisons were always identical and the remaining comparisons were different. For example, for the $P_3(x)$ comparisons we observe that the x^3, x^2 and x-coefficients from the polynomial $P_3(x)$ in (3.20), which was obtained from the generating function (3.5), were the same as those that arose out of computing $P_3(x)$ from the three-term recurrence relation (3.1); however, the constant terms were different. This pattern continued for the $P_4(x)$ and $P_5(x)$ comparisons and will continue for all other higher-order comparisons as well. This is apparent because the coefficients x^n, x^{n-1} and x^{n-2} obtained via the three-term recurrence relation (3.1) are dependent on A_n, B_n and C_n in (3.18), which were derived from the generating function (3.5), whereas the lower-order coefficients x^{n-k} for $k = 3, 4, \ldots, n-1$ and the constant terms are not.

Now that these relations have been derived, it is necessary to analyze them in order to make inferences on their nature so orthogonality conditions can be obtained. However, we first examine the complexities involved in the above relations in order to effectively determine additional assumptions that can simplify our calculations and reduce our problem to a manageable format. These details are discussed next.

3.6   Managing the Complexity of the Sheffer B-Type 1 Class

Here, we discuss the ramifications of the complexity of the *B-Type 1* polynomials $P_0(x), \ldots, P_5(x)$, which were determined in Sects. 3.4 and 3.5 and the comparisons obtained in Sect. 3.5. We emphasize that the coefficients, $c_{n,0}$, $c_{n,1}$ and $c_{n,2}$, as respectively defined in (3.15), (3.16) and (3.17), are *sums* involving the g_2-term and other multiplicative constants that grow arbitrary large as the polynomial degree increases. In fact, it can readily be shown that all of the coefficients of a given polynomial from the set $\{P_1(x), \ldots, P_5(x)\}$, and all higher-order polynomials as well, are also sums that grow arbitrary large as the polynomial degree increases. The notable exceptions to this are the constant terms, since for a given polynomial $P_k(x)$ the constant term is a_k. This is in stark contrast to the *B-Type 0* polynomials and is a crucial observation because the *B-Type 0* polynomials have coefficients that have a *fixed* structure.

To elaborate on this notion, we take $G(t) \equiv 0$ in (3.5) and obtain

$$A(t)\exp[xH(t)] = \sum_{k=0}^{\infty} P_k(x)t^n,$$

which is the generating function for the Sheffer *B-Type 0* polynomials as in (3.2). Also, we can obtain the coefficients of x^n, x^{n-1} and x^{n-2} for these polynomials by evaluating (3.15), (3.16) and (3.17) for $g_2 = g_3 = g_4 = 0$, since $G(t) \equiv 0$, resulting in

$$\check{c}_{n,0} := \frac{1}{n!} \tag{3.28}$$

$$\check{c}_{n,1} := \frac{a_1}{(n-1)!} + \frac{h_2}{(n-2)!} \tag{3.29}$$

$$\check{c}_{n,2} := \frac{a_2}{(n-2)!} + \frac{a_1 h_2 + h_3}{(n-3)!} + \frac{h_2^2}{2!(n-4)!}, \tag{3.30}$$

where we have labeled these coefficients $\check{c}_{n,0}$, $\check{c}_{n,1}$ and $\check{c}_{n,2}$ to distinguish them from the *B-Type 1* coefficients. It is now clear that for any n-value we have exactly one term in (3.28), at most two terms in (3.29) and at most three terms in (3.30). In contrast, (3.15) has exactly $\lfloor n/2 \rfloor + 1$ terms, (3.16) has exactly $\lfloor (n-1)/2 \rfloor + \lfloor (n-2)/2 \rfloor + \lfloor (n-3)/2 \rfloor + 3$ terms and (3.17) has 0, 1 and 3 terms for $n = 1, 2, 3$, respectively, and exactly $\lfloor (n-2)/2 \rfloor + 2(\lfloor (n-3)/2 \rfloor + 1) + 3(\lfloor (n-4)/2 \rfloor + 1) + \lfloor (n-5)/2 \rfloor + \lfloor (n-6)/2 \rfloor + 3$ terms for $n \geq 4$. Therefore, we obtain expressions for A_n, B_n, and C_n (which we label $\check{A}_n$, $\check{B}_n$ and $\check{C}_n$ for consistency) in the three-term recurrence relation (3.1) by substituting (3.28), (3.29) and (3.30) into (3.18):

$$\check{A}_n := \frac{1}{n+1},$$

$$\check{B}_n := \frac{a_1 + 2nh_2}{n+1},$$

$$\check{C}_n := \frac{1}{n+1}(a_1^2 - 2a_2 + 2a_1h_2 - 4h_2^2 + 3h_3 + (4h_2^2 - 3h_3)n).$$

Without any computations, it is intuitively obvious that the A_n, B_n, and C_n in the *B-Type 1* class will not be as simple to work with. For example, the expression for A_n was obtained using Mathematica® 4.1 as follows:

`Together[A[n]]`

$$\frac{-4\text{Gamma}[1+n]\text{HypergeometricU}[-\frac{1}{2}-\frac{n}{2},\frac{1}{2},-\frac{1}{4g_2}]g_2}{\text{Gamma}[2+n]\text{HypergeometricU}[\frac{1}{2}-\frac{n}{2},\frac{3}{2},-\frac{1}{4g_2}]},$$

with

$$\text{Gamma}[z] := \Gamma(z) = \int_0^\infty t^{z-1}e^{-t}dt, \qquad \text{Re}(z) > 0$$

$$\text{HypergeometricU}[a,b,z] := \Psi(a,b;z),$$

where Ψ denotes the Tricomi Ψ function, which is the second linearly independent solution to Kummer's differential equation: $zy'' + (b-z)y' - ay = 0$. It is defined by

$$\Psi(a,b;z) := \frac{\Gamma(1-b)}{\Gamma(a-b+1)}{}_1F_1(a,b;z) + \frac{\Gamma(b-1)}{\Gamma(a)}z^{1-b}{}_1F_1(a-b+1,2-b;z),$$

with ${}_1F_1(a,b;z)$ defined as

$$_1F_1(a,b,:z) := \sum_{k=0}^\infty \frac{(a)_k}{(b)_k}\frac{z^k}{k!}.$$

Remark 3.2. As we have mentioned, the expression for A[n] above was obtained using Mathematica® 4.1. Using Mathematica® 8, the output will be different, but nonetheless much more complicated than (3.28). Here, we displayed the Mathematica® 4.1. output since it was the most aesthetic.

Following a wealth of computing time, expressions were eventually obtained for B_n and C_n as well. However, these expressions are too large to display. For example, the actual size of the general expression for C_n exceeds two pages of Mathematica® output. Therefore, it is clear from simply studying the A_n term displayed above that the *B-Type 1* coefficients are significantly more difficult to work with than the *B-Type 0* coefficients.

We next address the very cumbersome expressions that arise when comparing respective coefficients from both the generating function (3.5) and the three-term recurrence relation (3.1) of a given *B-Type 1* polynomial as was done in Sect. 3.5. In particular, we see a wealth of g_3 and g_4 terms in the comparisons developed in Sect. 3.5—for a paradigm example, consider the $P_5(x)$ linear-term comparison (3.26). Therefore, the $P_4(x)$ and $P_5(x)$ comparisons (and any subsequent comparisons for that matter) would be much more manageable if we additionally take $g_i = 0, \forall i \geq 3$. We also mention that we want all simplifying assumptions to be restricted to alterations of only the $G(t)$ terms, as varying the terms of $H(t)$ or $A(t)$ would not reduce (3.5) to the *B-Type 0* class when we take $G(t) \equiv 0$.

Now, there are some very important structures embedded in the comparisons we have obtained and we address these next. We first turn our attention to the $P_3(x)$ constant-term comparison that was derived in (3.22), which we rewrite below with $g_i = 0, \forall i \geq 3$:

$$a_3 = \frac{1}{3(1+2g_2)^2}(-a_1^3 + 3a_1a_2 + 4a_1a_2g_2 - 12a_1^3g_2^2 + 12a_1a_2g_2^2$$

$$- 2a_1^2h_2 + 4a_2h_2 + 12a_1^2g_2h_2 - 4a_1h_2^2 + 3a_1h_3 + 12a_1g_2h_3 + 12a_1g_2^2h_3).$$

$$(3.31)$$

As seen above, and in all of the comparisons we established, g_2 is one of the most abundant terms, which is intuitively obvious since g_2 is in the argument of each sum in (3.15), (3.16) and (3.17). Therefore, it is natural to consider a simplifying assumption on g_2 that will reduce the complexity of (3.31) and therefore the remaining comparisons.

Now, several assumptions on the g_2-term were attempted before an appropriate choice was made. Below is an example of a particular choice for g_2 that could *not* be made but led to a beneficial result nonetheless. Notice that from setting (3.31) equal to zero, we have

$$- a_1^3 + 3a_1a_2 + 4a_1a_2g_2 - 12a_1^3g_2^2 + 12a_1a_2g_2^2 - 2a_1^2h_2 + 4a_2h_2$$

$$+ 12a_1^2g_2h_2 - 4a_1h_2^2 + 3a_1h_3 + 12a_1g_2h_3 + 12a_1g_2^2h_3 - 3a_3(1+2g_2)^2 = 0.$$

Then, the choice of $g_2 = -1/2$ would reduce the above to the following format:

$$-a_1^3 + a_1a_2 - 2a_1^2h_2 + a_2h_2 - a_1h_2^2 = 0.$$

However, the choice of $g_2 = -1/2$ results in (3.19) becoming

$$P_2(x) = a_2 + (a_1 + h_2)x$$

since $g_2 = -1/2$ is a zero of (3.15) for $n = 2$. This is of course not permissible as there would never be a polynomial of degree 2 in the sequence $\{P_n(x)\}_{n=0}^{\infty}$ as defined by (3.5).

Nonetheless, we now consider the choice of $g_2 = 1/2$, which results in a modified expression for a_3 in (3.31) as seen below

$$a_3 = \frac{1}{3}(-a_1^3 + 2a_1a_2 + a_1^2h_2 + a_2h_2 - a_1h_2^2 + 3a_1h_3).$$ (3.32)

Thus, in weighing (3.32) against (3.22), the discrepancies in the levels of complexity are apparent.

Now that all of our simplifying assumptions have been finalized, we restate our current problem. The generating function (3.5) now becomes

$$A(t)\exp\left[xH(t) + \frac{1}{2}x^2t^2\right] = \sum_{n=0}^{\infty} P_n(x)t^n$$ (3.33)

and (3.15), (3.16) and (3.17) now, respectively, have the form

$$c_{n,0} := \phi_n(1/2),$$
$$c_{n,1} := a_1\phi_{n-1}(1/2) + h_2\phi_{n-2}(1/2),$$ (3.34)
$$c_{n,2} := a_2\phi_{n-2}(1/2) + (h_3 + a_1h_2)\phi_{n-3}(1/2) + (h_2^2/2!)\phi_{n-4}(1/2).$$

We next determine what the assumptions $g_2 = 1/2$ and $g_i = 0, \forall i \geq 3$ do to the later comparisons developed in Sect. 3.5, starting with the constant- and linear-term comparisons of $P_4(x)$ and continuing through the constant, linear and quadratic-term comparisons of $P_5(x)$.

The $P_4(x)$ constant-term comparison (3.24) now becomes

$$a_4 = \frac{1}{16}(-a_1^2a_2 + 6a_2^2 + a_1a_3 - 4a_1a_2h_2 + 9a_3h_2 - a_2h_2^2 + 6a_2h_3)$$ (3.35)

and the $P_4(x)$ linear-term comparison (3.23) simplifies to

$$-a_1^3 + 7a_1a_2 - 6a_3 - 4a_1^2h_2 - a_2h_2 + 4a_1h_2^2 - h_2^3 - 9a_1h_3$$
$$+ 15h_2h_3 - 16h_4 = 0.$$ (3.36)

Then, for the $P_5(x)$ constant-term comparison, we see that (3.27) takes the form

$$a_5 = \frac{1}{250}(-42a_1^2a_3 + 55a_2a_3 + 42a_1a_4 + 29a_1a_3h_2 + 88a_4h_2$$
$$- 42a_3h_2^2 + 180a_3h_3),$$ (3.37)

the $P_5(x)$ linear-term comparison (3.26) becomes

$$- 42a_1^2 a_2 + 55a_2^2 + 42a_1 a_3 - 120a_4 - 42a_1^3 h_2 + 126a_1 a_2 h_2 - 162a_3 h_2$$

$$+ 29a_1^2 h_2^2 + 46a_2 h_2^2 - 42a_1 h_2^3 - 15a_2 h_3 + 297a_1 h_2 h_3 - 42h_2^2 h_3 + 180h_3^2$$

$$- 208a_1 h_4 + 88h_2 h_4 - 250h_5 = 0 \tag{3.38}$$

and the $P_5(x)$ quadratic-term comparison (3.25) is now

$$- 42a_1^3 + 97a_1 a_2 - 120a_3 + 29a_1^2 h_2 + 23a_2 h_2 - 29a_1 h_2^2 + 2h_2^3$$

$$+ 102a_1 h_3 + 18h_2 h_3 - 120h_4 = 0. \tag{3.39}$$

In all of the comparisons above, the reduction in complexity is evident.

Now that the above comparisons have been established, we see some very important potential patterns. First, we notice that the highest h-term in the expression for a_3 as defined by (3.32) is h_3—this is also the case for the expressions for a_4 in (3.35) and a_5 in (3.37). Second, we observe that the highest h-term in the $P_4(x)$ linear-term comparison (3.36) is h_4 and the highest h-term in the $P_5(x)$ linear-term comparison (3.38) is h_5. These observations are important because they provide information on how to develop conditions that the terms a_1, a_2, a_3 and h_2 must satisfy in order for $\{P_n(x)\}$ in (3.33) to be orthogonal. We discuss this further.

Based on the above analysis, we first find an expression for h_3 by utilizing the relation for a_3 in (3.32), which results in

$$h_3 = \frac{1}{3a_1}(a_1^3 + 3a_3 - 2a_1 a_2 - a_1^2 h_2 - a_2 h_2 + a_1 h_2^2), \quad a_1 \neq 0. \tag{3.40}$$

Therefore, we have written h_3 in terms of only a_1, a_2, a_3 and h_2. Now notice that the $P_4(x)$ linear-term comparison (3.36) can be solved for h_4 as follows

$$h_4 = \frac{1}{16}\left(-a_1^3 + 7a_1 a_2 - 6a_3 - 4a_1^2 h_2 - a_2 h_2 + 4a_1 h_2^2 - h_2^3 - 9a_1 h_3 + 15h_2 h_3\right). \tag{3.41}$$

Moreover, we can then substitute (3.40) into (3.41) and after some algebra we obtain an expression for h_4 that involves only a_1, a_2, a_3 and h_2, as written below

$$h_4 = -\frac{1}{16a_1}(a_1 - h_2)\left(4a_1^3 + 15a_3 - 5a_2 h_2 + a_1\left(-13a_2 + 4h_2^2\right)\right). \tag{3.42}$$

Lastly, upon substituting (3.40) and (3.42) into (3.39) and using some algebraic manipulations we see that the $P_5(x)$ quadratic-term comparison becomes

$$\frac{1}{a_1}(a_1 - h_2)\left(44a_1^3 - 14a_1^2 h_2 - 63\left(-3a_3 + a_2 h_2\right) + a_1\left(-137a_2 + 44h_2^2\right)\right) = 0. \tag{3.43}$$

Thus, we have obtained the algebraic equation (3.43) that relates all of the terms a_1, a_2, a_3, and h_2. Therefore, since the comparisons of the coefficients of the polynomials $P_6(x)$, $P_7(x)$ and $P_8(x)$ must behave in a similar fashion as the comparisons of the coefficients of the polynomials $P_3(x)$, $P_4(x)$ and $P_5(x)$ (which we will show), we can construct a system of four simultaneous nonlinear equations with unknowns a_1, a_2, a_3 and h_2. Solving this system will yield the conditions that $\{P_n(x)\}_{n=0}^{\infty}$ as defined by (3.33) must satisfy in order to be an orthogonal sequence.

Based on the analysis that led to (3.43), we now derive comparisons using the generating function (3.33) and the three-term recurrence relation (3.1), with the recursion coefficients as defined in (3.18) with the leading coefficients as in (3.34), for the $P_6(x)$ cubic term, the $P_7(x)$ fourth-degree term and the $P_8(x)$ fifth-degree term. We do this since from our previous analysis it appears that each of these comparisons will contain the terms a_1, a_2 and a_3 and the highest h-term will be h_4. Therefore, we can derive three more equations analogous to (3.43) by using the same methodology that led to (3.43). These computations are as follows and each of which was completed using the same procedure as in Sect. 3.4 for determining the polynomials $P_2(x), \ldots, P_5(x)$. For these results we initially keep h_3 and h_4 arbitrary.

The $P_6(x)$ polynomial that results from expanding (3.33) accordingly is

$$
\begin{aligned}
P_6(x) =\; & a_6 + (a_5 + a_4 h_2 + a_3 h_3 + a_2 h_4 + a_1 h_5 + h_6)x \\
& + \frac{1}{2}\left(2a_4 + 2a_3 h_2 + a_2 h_2^2 + 2a_2 h_3 + 2a_1 h_2 h_3 + h_3^2 + 2a_1 h_4 + 2h_2 h_4 + 2h_5\right)x^2 \\
& + \frac{1}{6}(4a_3 + 6a_2 h_2 + 3a_1 h_2^2 + h_2^3 + 6a_1 h_3 + 6h_2 h_3 + 6h_4)x^3 \\
& + \frac{1}{12}(5a_2 + 8a_1 h_2 + 6h_2^2 + 8h_3)x^4 \\
& + \frac{1}{60}(13a_1 + 25h_2)x^5 + \frac{19}{180}x^6,
\end{aligned}
$$

the $P_7(x)$ polynomial that comes from (3.33) is found to be

$$
\begin{aligned}
P_7(x) =\; & a_7 + (a_6 + a_5 h_2 + a_4 h_3 + a_3 h_4 + a_2 h_5 + a_1 h_6 + h_7)x \\
& + \frac{1}{2}\Big(2a_5 + 2a_4 h_2 + a_3 h_2^2 + 2a_3 h_3 + 2a_2 h_2 h_3 + a_1 h_3^2 + 2a_2 h_4 \\
& \qquad + 2a_1 h_2 h_4 + 2h_3 h_4 + 2a_1 h_5 + 2h_2 h_5 + 2h_6\Big)x^2 \\
& + \frac{1}{6}\Big(4a_4 + 6a_3 h_2 + 3a_2 h_2^2 + a_1 h_2^3 + 6a_2 h_3 + 6a_1 h_2 h_3 + 3h_2^2 h_3 \\
& \qquad + 3h_3^2 + 6a_1 h_4 + 6h_2 h_4 + 6h_5\Big)x^3 \\
& + \frac{1}{12}\left(5a_3 + 8a_2 h_2 + 6a_1 h_2^2 + 2h_2^3 + 8a_1 h_3 + 12h_2 h_3 + 8h_4\right)x^4
\end{aligned}
$$

$$+ \frac{1}{60} \left(13a_2 + 25a_1h_2 + 20h_2^2 + 25h_3 \right) x^5$$

$$+ \frac{1}{180} \left(19a_1 + 39h_2 \right) x^6 + \frac{29}{630} x^7$$

and for $P_8(x)$ we have the following:

$$P_8(x) = a_8 + \left(a_7 + a_6h_2 + a_5h_3 + a_4h_4 + a_3h_5 + a_2h_6 + a_1h_7 + h_8 \right) x$$

$$+ \frac{1}{2} \left(2a_6 + 2a_5h_2 + a_4h_2^2 + 2a_4h_3 + 2a_3h_2h_3 + a_2h_3^2 + 2a_3h_4 + 2a_2h_2h_4 \right.$$

$$+ 2a_1h_3h_4 + h_4^2 + 2a_2h_5 + 2a_1h_2h_5 + 2h_3h_5 + 2a_1h_6 + 2h_2h_6 + 2h_7 \left. \right) x^2$$

$$+ \frac{1}{6} \left(4a_5 + 6a_4h_2 + 3a_3h_2^2 + a_2h_2^3 + 6a_3h_3 + 6a_2h_2h_3 + 3a_1h_2^2h_3 + 3a_1h_3^2 \right.$$

$$+ 3h_2h_3^2 + 6a_2h_4 + 6a_1h_2h_4 + 3h_2^2h_4 + 6h_3h_4 + 6a_1h_5 + 6h_2h_5 + 6h_6 \left. \right) x^3$$

$$+ \frac{1}{24} \left(10a_4 + 16a_3h_2 + 12a_2h_2^2 + 4a_1h_2^3 + h_2^4 + 16a_2h_3 + 24a_1h_2h_3 \right.$$

$$+ 12h_2^2h_3 + 12h_3^2 + 16a_1h_4 + 24h_2h_4 + 16h_5 \left. \right) x^4$$

$$+ \frac{1}{60} \left(13a_3 + 25a_2h_2 + 20a_1h_2^2 + 10h_2^3 + 25a_1h_3 + 40h_2h_3 + 25h_4 \right) x^5$$

$$+ \frac{1}{360} \left(38a_2 + 78a_1h_2 + 75h_2^2 + 78h_3 \right) x^6$$

$$+ \frac{1}{1260} \left(58a_1 + 133h_2 \right) x^7 + \frac{191}{10080} x^8.$$

By using the same method as conducted in Sect. 3.5, we now derive the $P_6(x)$ cubic-term comparison, the $P_7(x)$ fourth-degree term comparison and the $P_8(x)$ fifth-degree term comparison. We first state each respective comparison and then substitute the expressions for h_3 and h_4, as respectively in (3.40) and (3.42), using a wealth of algebraic manipulations in the process. The $P_6(x)$ cubic-term comparison is

$$- 320a_1^3 + 1438a_1a_2 - 1365a_3 - 478a_1^2h_2 - 73a_2h_2 + 478a_1h_2^2 - 135h_2^3$$

$$- 1236a_1h_3 + 2601h_2h_3 - 3900h_4 = 0. \tag{3.44}$$

Upon substituting (3.40) and (3.42) into (3.44) we obtain

$$\frac{1}{a_1} \left(a_1 - h_2 \right) \left(324a_1^3 + 92a_1^2h_2 - 469 \left(-3a_3 + a_2h_2 \right) + a_1 \left(-1209a_2 + 324h_2^2 \right) \right) = 0. \tag{3.45}$$

The $P_7(x)$ fourth-degree term comparison becomes

$$-17225a_1^3 + 43730a_1a_2 - 49647a_3 + 7945a_1^2h_2 + 5917a_2h_2 - 7945a_1h_2^2$$
$$+676h_2^3 + 32139a_1h_3 + 17508h_2h_3 - 90896h_4 = 0 \qquad (3.46)$$

and after substituting (3.40) and (3.42) into (3.46) we have

$$\frac{1}{a_1}(a_1 - h_2)\left(5404a_1^3 - 1148a_1^2h_2 - 7523\left(-3a_3 + a_2h_2\right)\right)$$
$$+a_1\left(-17183a_2 + 5404h_2^2\right)) = 0. \qquad (3.47)$$

Lastly, the $P_8(x)$ fifth-degree term comparison is as follows:

$$-372723a_1^3 + 1403441a_1a_2 - 1385214a_3 - 285272a_1^2h_2 - 18227a_2h_2$$
$$+285272a_1h_2^2 - 89015h_2^3 - 803811a_1h_3 + 2189025h_2h_3 - 4363920h_4 = 0. \qquad (3.48)$$

From substituting (3.40) and (3.42) into (3.48) we observe that

$$\frac{1}{a_1}(a_1 - h_2)\left(45032a_1^3 + 7168a_1^2h_2 - 63405\left(-3a_3 + a_2h_2\right)\right)$$
$$+a_1\left(-160637a_2 + 45032h_2^2\right)) = 0. \qquad (3.49)$$

Hence, (3.43), (3.45), (3.47) and (3.49) result in a simultaneous system of nonlinear algebraic equations in the variables a_1, a_2, a_3 and h_2 as seen below:

$$\left.\begin{array}{c} \frac{1}{a_1}(a_1 - h_2)\left(44a_1^3 - 14a_1^2h_2 - 63\left(-3a_3 + a_2h_2\right)\right) \\ +a_1\left(-137a_2 + 44h_2^2\right)) = 0 \\ \frac{1}{a_1}(a_1 - h_2)\left(324a_1^3 + 92a_1^2h_2 - 469\left(-3a_3 + a_2h_2\right)\right) \\ +a_1\left(-1209a_2 + 324h_2^2\right)) = 0 \\ \frac{1}{a_1}(a_1 - h_2)\left(5404a_1^3 - 1148a_1^2h_2 - 7523\left(-3a_3 + a_2h_2\right)\right) \\ +a_1\left(-17183a_2 + 5404h_2^2\right)) = 0 \\ \frac{1}{a_1}(a_1 - h_2)\left(45032a_1^3 + 7168a_1^2h_2 - 63405\left(-3a_3 + a_2h_2\right)\right) \\ +a_1\left(-160637a_2 + 45032h_2^2\right)) = 0 \end{array}\right\}. \qquad (3.50)$$

3.7 The Solutions of the Simultaneous, Nonlinear, Algebraic System

If all of the solutions to (3.50) are found, exactly one of two situations will transpire. One, at least one of the solutions will yield restrictions on any number of the terms a_1, a_2, a_3 and h_2, satisfying both (3.33) and (3.1), i.e., (3.33) will produce at least one orthogonal polynomial sequence. Two, each of the solutions will lead to a contradiction. In this case, it would be shown that $\{P_n(x)\}_{n=0}^{\infty}$ as defined by (3.33) does not yield any orthogonal polynomial sequences. In the analysis that follows, we illustrate that the latter case applies by showing that each solution set to (3.50) contains the requirement $a_1 = h_2$, which leads to a contradiction. We begin with the formal statement of this below.

Theorem 3.3. *Each solution set to the simultaneous, nonlinear, algebraic system defined by (3.50), which was established by respectively comparing the $P_5(x)$ quadratic, $P_6(x)$ cubic, $P_7(x)$ quartic and $P_8(x)$ quintic coefficients from the three-term recurrence relation (3.1) (for A_n, B_n and C_n as in (3.18) with the leading coefficients as defined by (3.34)) with the $P_5(x)$ quadratic, $P_6(x)$ cubic, $P_7(x)$ quartic and $P_8(x)$ quintic coefficients from the generating function (3.33), contains the requirement $a_1 = h_2$.*

Proof. Notice that each equation in (3.50) has the following structure:

$$d_1 a_1^2 \delta + d_2 a_1 h_2 \delta + d_3(-3a_3 + a_2 h_2)\delta/a_1 + d_4 a_2 \delta + d_1 h_2^2 \delta = 0,$$

where $\delta = a_1 - h_2$. Therefore, we consider a linear change-of-variables by assigning $A = a_1^2 \delta$, $B = a_1 h_2 \delta$, $C = (-3a_3 + a_2 h_2)\delta/a_1$, $D = a_2 \delta$ and $E = h_2^2 \delta$. Then, (3.50) can be written as

$$\begin{bmatrix} 44 & -14 & -63 & -137 & 44 \\ 324 & 92 & -469 & -1209 & 324 \\ 5404 & -1148 & -7523 & -17183 & 5404 \\ 45032 & 7168 & -63405 & -160637 & 45032 \end{bmatrix} \begin{bmatrix} A \\ B \\ C \\ D \\ E \end{bmatrix} = \begin{bmatrix} 0 \\ 0 \\ 0 \\ 0 \end{bmatrix}.$$

Elementary row reductions to the above system lead to:

$$\begin{bmatrix} 1 & 0 & 0 & -2 & 1 & 0 \\ 0 & 1 & 0 & -1 & 0 & 0 \\ 0 & 0 & 1 & 1 & 0 & 0 \\ 0 & 0 & 0 & 0 & 0 & 0 \end{bmatrix},$$

which implies that $C = -D$, $B = D$ and $A = 2D - E$.

Now, we clearly see that $\delta = 0$ ($a_1 = h_2$) implies that $A = B = C = D = E = 0$, which satisfies the above requirements and trivially solves the above system. Therefore, we next omit the δ-factor from each of the assignments $A, \ldots, E$ above. This yields $-3a_3 + a_2 h_2 = -a_1 a_2$, $a_1 h_2 = a_2$ and $a_1^2 = 2a_2 - h_2^2$, where the last two equations lead to $a_1^2 = 2a_1 h_2 - h_2^2$, which clearly has only the solution $a_1 = h_2$ (multiplicity 2). Hence, we see that every solution set of (3.50) that is additional to $\delta = 0$ has $a_1 = h_2$ as a requirement as well and the theorem is proven.   □

We now show that the condition $a_1 = h_2$ leads to a contradiction by constructing another simultaneous system. First, recall that the highest h-term in the $P_5(x)$ linear-term comparison (3.38) is h_5. In addition, the highest h-term in each of the $P_6(x)$ quadratic-term comparison, the $P_7(x)$ cubic-term comparison and the $P_8(x)$ fourth-degree term comparison is also h_5. These comparisons are listed below, under the assumption $a_1 = h_2$.

The $P_6(x)$ quadratic-term comparison is

$$-86a_2^2 + 200a_4 + 10a_3h_2 - 38a_2h_2^2 + 34h_2^4$$
$$-54a_2h_3 - 156h_2^2h_3 + 27h_3^2 + 400h_2h_4 + 200h_5 = 0, \qquad (3.51)$$

the $P_7(x)$ cubic-term comparison turns out to be

$$-1116a_2^2 + 2392a_4 + 2834a_3h_2 - 2994a_2h_2^2 + 1224h_2^4$$
$$+156a_2h_3 - 5382h_2^2h_3 - 3105h_3^2 + 7176h_2h_4 + 5850h_5 = 0 \qquad (3.52)$$

and the $P_8(x)$ fourth-degree term comparison is

$$-6835a_2^2 + 15048a_4 + 8835a_3h_2 - 10213a_2h_2^2 + 4888h_2^4$$
$$-28111a_2h_3 - 21072h_2^2h_3 + 954h_3^2 + 60192h_2h_4 + 26980h_5 = 0. \qquad (3.53)$$

Each of the above comparisons (3.51), (3.52) and (3.53) were all obtained using the same method that has been utilized throughout this chapter.

Now, we note that under the restriction $a_1 = h_2$, (3.40) and (3.42), respectively, take on the form

$$h_3 = \frac{1}{3h_2}(3a_3 - 3a_2h_2 + h_2^3), \quad h_2 \neq 0 \quad \text{and} \quad h_4 = 0. \qquad (3.54)$$

Also, we note that (3.51), (3.52) and (3.53) all have an a_4 term and upon substituting h_3 as in (3.54) and incorporating the requirement $a_1 = h_2$, our a_4 expression in (3.35) now becomes

$$a_4 = \frac{1}{8h_2}\left(3a_2a_3 + 5a_3h_2^2 - 2a_2h_2^3\right). \qquad (3.55)$$

Therefore, we can substitute (3.54) and (3.55) into (3.38) and solve for h_5 yielding

$$h_5 = \frac{1}{25h_2^2}(18a_3^2 - 42a_2a_3h_2 + 25a_2^2h_2^2 + 18a_3h_2^3 - 22a_2h_2^4 + 5h_2^6). \qquad (3.56)$$

Thus, our expression for h_5 involves only the variables a_2, a_3 and h_2. Then, substituting (3.54), (3.55) and (3.56) into (3.51), (3.52) and (3.53) and using some algebraic manipulations, we construct a simultaneous system in the variables a_2, a_3 and h_2, as seen below:

$$\left. \begin{array}{l} \left(171a_3^2 + 195a_2^2h_2^2 - 144a_2h_2^4 \right. \\ \left. \quad + 25h_2^6 + a_3\left(-369a_2h_2 + 141h_2^3\right)\right) h_2^{-2} = 0 \\ \left(369a_3^2 + 491a_2^2h_2^2 - 412a_2h_2^4 \right. \\ \left. \quad + 85h_2^6 + a_3\left(-855a_2h_2 + 363h_2^3\right)\right) h_2^{-2} = 0 \\ \left(16983a_3^2 + 19925a_2^2h_2^2 - 15182a_2h_2^4 \right. \\ \left. \quad + 2805h_2^6 + a_3\left(-37002a_2h_2 + 14358h_2^3\right)\right) h_2^{-2} = 0 \end{array} \right\}. \tag{3.57}$$

To solve (3.57) we first note that each equation has the following structure:

$$\eta_1 a_3^2 h_2^{-2} + \eta_2 a_2^2 + \eta_3 a_2 h_2^2 + \eta_4 h_2^4 + \eta_5 a_2 a_3 h_2^{-1} + \eta_6 a_3 h_2 = 0.$$

Thus, we again use a linear change-of-variables by assigning $A = a_3^2 h_2^{-2}, B = a_2^2, C = a_2 h_2^2, D = h_2^4, E = a_2 a_3 h_2^{-1}$ and $F = a_3 h_2$. This leads to the linear system

$$\begin{bmatrix} 171 & 195 & -144 & 25 & -369 & 141 \\ 369 & 491 & -412 & 85 & -855 & 363 \\ 16983 & 19925 & -15182 & 2805 & -37002 & 14358 \end{bmatrix} \begin{bmatrix} A \\ B \\ C \\ D \\ E \\ F \end{bmatrix} = \begin{bmatrix} 0 \\ 0 \\ 0 \end{bmatrix},$$

which has the following reduced echelon form:

$$\begin{bmatrix} 1 & 0 & 0 & 4/9 & 0 & -4/3 & 0 \\ 0 & 1 & 0 & -1 & -3 & 3 & 0 \\ 0 & 0 & 1 & -1 & -3/2 & 3/2 & 0 \end{bmatrix}.$$

Therefore, we have $9A = 12F - 4D, B = D + 3(E - F)$ and $2C = B + D$, which is of course equivalent to

$$\left. \begin{array}{l} 9a_3^2 h_2^{-2} = 12a_3h_2 - 4h_2^4 \\ a_2^2 = h_2^4 + 3a_3(a_2h_2^{-1} - h_2) \\ 2a_2h_2^2 = a_2^2 + h_2^4 \end{array} \right\}. \tag{3.58}$$

Then, from the bottom equation in (3.58), we see that $a_2 = h_2^2$, which clearly satisfies the middle equation. Lastly, the top equation gives $a_3 = 2h_2^3/3$ (multiplicity 2) and therefore, $\left\{a_3 = 2h_2^3/3, \ a_2 = h_2^2\right\}$ solves (3.57). Upon extracting all possible solutions to (3.58) we prove the following statement.

Theorem 3.4. *The simultaneous nonlinear algebraic system defined by (3.57) has only the solution sets*

$$\left\{ a_2 = \left(\pm \frac{3}{2} \right)^{2/3} a_3^{2/3}, \quad h_2 = \pm \left(\pm \frac{3}{2} \right)^{1/3} a_3^{1/3} \right\},$$

$$\left\{ a_3 = \frac{2}{3} h_2^3, \quad a_2 = h_2^2 \right\},$$

$$\left\{ a_2 = -(-1)^{1/3} \left(\frac{3}{2} \right)^{2/3} a_3^{2/3}, \quad h_2 = (-1)^{2/3} \left(\frac{3}{2} \right)^{1/3} a_3^{1/3} \right\},$$

$$\left\{ a_3 = \pm \frac{2}{3} a_2^{3/2}, \quad h_2 = \pm \sqrt{a_2} \right\}.$$

Now consider $A_2 A_1 C_2$, with $a_1 = h_2$. With some algebraic manipulations we have

$$A_2 A_1 C_2 = \frac{4a_2}{9} - \frac{2a_3}{3h_2}. \tag{3.59}$$

With the restrictions $a_3 = 2h_2^3/3$ and $a_2 = h_2^2$ from the third solution set of Theorem 3.4, we clearly see that (3.59) becomes $A_2 A_1 C_2 = 0$, which violates the positivity condition $A_n A_{n-1} C_n > 0$ in (3.1) for $n = 2$. It can be verified by simple substitutions that *all* of the other solution sets also make (3.59) equal to zero, thus violating the positivity condition in (3.1) for $n = 2$.

It is very important to discuss the fact that in order to construct the system (3.50) we assumed $a_1 \neq 0$ via (3.40). Thus, to complete our investigation we need to additionally consider the case when $a_1 = 0$. Therefore, assuming $a_1 = 0$, the $P_5(x)$ quadratic-term comparison (3.39) is

$$-120a_3 + 23a_2 h_2 + 2h_2^3 + 18h_2 h_3 - 120h_4 = 0, \tag{3.60}$$

the $P_6(x)$ cubic-term comparison (3.44) becomes

$$-1365a_3 - 73a_2 h_2 - 135h_2^3 + 2601h_2 h_3 - 3900h_4 = 0 \tag{3.61}$$

and the $P_7(x)$ fourth-degree term comparison (3.46) transforms into

$$-49647a_3 + 5917a_2 h_2 + 676h_2^3 + 17508h_2 h_3 - 90896h_4 = 0. \tag{3.62}$$

In addition, our expression for a_3 in (3.32) becomes

$$a_3 = \frac{1}{3} a_2 h_2 \tag{3.63}$$

and h_4 in (3.41) turns into

$$h_4 = \frac{1}{16}(-6a_3 - a_2 h_2 - h_2^3 + 15 h_2 h_3).$$

Then, substituting (3.63) into the h_4 expression above yields

$$h_4 = \frac{1}{16}(-3a_2 h_2 - h_2^3 + 15 h_2 h_3). \tag{3.64}$$

Next, upon substituting (3.63) and (3.64) into each of the equations (3.60), (3.61) and (3.62) we achieve:

$$\left.\begin{array}{l} h_2\left(11a_2 + 19h_2^2 - 189h_3\right) = 0 \\ h_2\left(271a_2 + 145h_2^2 - 1407h_3\right) = 0 \\ h_2\left(2137a_2 + 2119h_2^2 - 22569h_3\right) = 0 \end{array}\right\}. \tag{3.65}$$

It is not difficult to show that the only solution sets to (3.65) are $\{h_2 = 0\}$ and $\{a_2 = h_2 = h_3 = 0\}$. We first assume $h_2 = 0$ and immediately see that under this assumption, $a_3 = h_4 = 0$ via (3.63) and (3.64). Therefore, we take the restrictions $a_1 = a_3 = h_2 = h_4 = 0$ and substitute them into the $P_6(x)$ quadratic-term comparison (3.51) and our $P_7(x)$ cubic-term comparison (3.52), which respectively yields

$$-86a_2^2 + 200a_4 - 54a_2 h_3 + 27h_3^2 + 200h_5 = 0 \tag{3.66}$$

and

$$-1116a_2^2 + 2392a_4 + 156a_2 h_3 - 3105h_3^2 + 5850h_5 = 0. \tag{3.67}$$

Substituting $a_1 = a_3 = h_2 = h_4 = 0$ into the a_4 relation (3.35) and the $P_5(x)$ linear-term comparison (3.38) and solving the latter for h_5 we obtain

$$a_4 = \frac{3a_2}{8}(a_2 + h_3) \quad \text{and} \quad h_5 = \frac{1}{50}(11a_2^2 - 24a_4 - 3a_2 h_3 + 36h_3^2). \tag{3.68}$$

Then substituting a_4 above into h_5 above leads to

$$h_5 = \frac{1}{25}\left(a_2^2 - 6a_2 h_3 + 18h_3^2\right). \tag{3.69}$$

Therefore, substituting a_4 in (3.68) and h_5 in (3.69) into both (3.66) and (3.67) gives

$$\left.\begin{array}{l} a_2^2 + 9a_2 h_3 - 57h_3^2 = 0 \\ 5a_2^2 - 117a_2 h_3 + 369h_3^2 = 0 \end{array}\right\}. \tag{3.70}$$

It can readily be shown that the only solution to the system (3.70) is $a_2 = h_3 = 0$. From here, it can easily be verified that the restriction $a_2 = h_3 = 0$ coupled with the restrictions $a_1 = a_3 = h_2 = h_4 = 0$ again leads to a contradiction of the positivity condition $A_n A_{n-1} C_n > 0$ in (3.1) for $n = 2$. Lastly, it can be shown by direct substitution of the constraints in the solution set $\{a_2 = h_2 = h_3 = 0\}$ of (3.65), along with the requirement $a_1 = 0$, that we once again have a contradiction of the positivity condition $A_n A_{n-1} C_n > 0$ in (3.1) for $n = 2$.

To summarize, for the sequence $\{P_n(x)\}_{n=0}^{\infty}$ in (3.33) to be orthogonal, the system defined by (3.50) must have at least one solution that does not lead to a contradiction. As we have shown, every solution to the system leads to a violation of the positivity condition of (3.1). Furthermore, since the system (3.50) was constructed under the assumption $a_1 \neq 0$, we additionally considered the case when $a_1 = 0$ and constructed another simultaneous system, the only solutions of which each led to a contradiction as well. Thus, we have now exhausted all possibilities and in turn have established the following statement.

Theorem 3.5. *There exists no sequences $\{a_i\}_{i=0}^{\infty}$ and $\{h_i\}_{i=1}^{\infty}$ such that $\{P_n(x)\}_{n=0}^{\infty} \equiv \{Q_n(x)\}_{n=0}^{\infty}$, with $\{P_n(x)\}_{n=0}^{\infty}$ defined by*

$$A(t)\exp\left[xH(t) + \frac{1}{2}x^2 t^2\right] = \sum_{n=0}^{\infty} P_n(x)t^n$$

with $A(t) = \sum_{i=0}^{\infty} a_i t^i$, $a_0 = 1$ and $H(t) = \sum_{i=1}^{\infty} h_i t^i$, $h_1 = 1$

and $\{Q_n(x)\}_{n=0}^{\infty}$ defined by

$$Q_{n+1}(x) = (A_n x + B_n)Q_n(x) - C_n Q_{n-1}(x), \quad A_n A_{n-1} C_n > 0,$$

$$\text{where } Q_{-1}(x) = 0 \text{ and } Q_0(x) = 1.$$

Or simply, there exist no orthogonal polynomial sequences $\{P_n(x)\}_{n=0}^{\infty}$ that satisfy (3.33).

3.8   On the Verification of Solutions via Computer Algebra

Herein, we give an overview on how two Mathematica® commands (Solve and Reduce), which have the ability to potentially solve simultaneous nonlinear algebraic systems, aided in the development of Theorems 3.3 and 3.4. The first Mathematica® command is entitled Solve, which yields only generic solutions, i.e., conditions on the variables that one explicitly solves for and not on any other parameters in the system—refer to [5] for more details.

Now, even though the system (3.50) can be potentially solved by Mathematica® it is imperative to explain how this is accomplished, as there are some subtleties. In fact, any combination of the terms a_1, a_2, a_3 and h_2 can be treated as variables and in addition, no variables can be specified, and then Mathematica® can attempt to find explicit solutions as well as implicit solutions. Therefore, to more accurately determine the solutions to the system (3.50), we must exhaust all possible variable selections and there are of course a total of $\sum_{n=1}^{4} C(4,n) + 1 = 16$ choices. All possible variable selections were tested and each yielded $a_1 = h_2$ as a requirement in each solution set. The procedure was carried out as follows.

We first assign the left-hand sides of the equations in (3.50) as $E1, \ldots, E4$, set each equal to zero and then have Mathematica® search for all possible solutions with respect to each selection of parameters. As an example, below we display five of the 16 total outputs:

In[1]:=Solve[{E1==0, E2==0, E3==0, E4==0},{ a_1 }]
Out[1]={{$h_2 \to a_1$}}

In[2]:=Solve[{E1==0, E2==0, E3==0, E4==0},{ a_2 }]
Out[2]={}

In[3]:=Solve[{E1==0, E2==0, E3==0, E4==0},{ a_3 }]
Out[3]={}

In[4]:=Solve[{E1==0, E2==0, E3==0, E4==0},{ h_2 }]
Out[4]={{$h_2 \to a_1$}}

In[5]:=Solve[{E1==0, E2==0, E3==0, E4==0},{ a_1,a_2,a_3,h_2 }]
Solve::svars : Equations may not give solutions for all ``solve''
variables.

$$Out[5] = \left\{ \left\{ a_3 \to \frac{-2(48616a_1^3 - 112021a_1a_2)}{190215}, h_2 \to a_1 \right\}, \right.$$
$$\left\{ a_3 \to \frac{-2(4830a_1^3 - 123531a_1a_2)}{22569}, h_2 \to a_1 \right\},$$
$$\left\{ a_3 \to \frac{-2(370a_1^3 - 8391a_1a_2)}{1407}, h_2 \to a_1 \right\},$$
$$\left\{ a_3 \to \frac{-2(37a_1^3 - 100a_1a_2)}{189}, h_2 \to a_1 \right\},$$
$$\left\{ a_3 \to \frac{2a_1^3}{3}, a_2 \to a_1^2, h_2 \to a_1 \right\},$$
$$\left. \{a_1 \to h_2\} \right\}.$$

We first note that the solution $\left\{ a_3 \to \frac{2a_1^3}{3}, a_2 \to a_1^2, h_2 \to a_1 \right\}$ in Output 5 is listed 12 times in the actual Mathematica® file, because Mathematica® calculates its multiplicity to be 12. Also in Output 5, we note that all of the solutions contain the assignment $a_1 = h_2$.

Altogether, each variable selection either yielded a null output, a solution that required $a_1 = h_2$ or a solution set that can be algebraically manipulated to yield

$a_1 = h_2$. For example, one of the solutions for the variable selection $\{a_1, a_3, h_2\}$ was determined to be

$$\left\{ a_3 \to -2a_2^{3/2}/3, \ h_2 \to -\sqrt{a_2}, \ a_1 \to -\sqrt{a_2} \right\}$$

from which it is readily seen that $a_1 = h_2$. This leads to the conjecture that *each* solution set of (3.50) contains the requirement $a_1 = h_2$.

We additionally mention that in Output 5 we see that Mathematica® displays the preface; *"Equations may not give solutions for all "solve" variables,"* which indicates that Mathematica® did not necessarily solve the system, so the `Solve` command cannot be interpreted as definitive. Therefore, as an additional verification, we repeat what has been done above using a different command—the `Reduce`. This command does not disregard restrictions on any other parameters in the solutions, as opposed to the `Solve` command. Moreover, the `Reduce` command also disregards multiplicities of solutions as it always displays each solution set only once—refer to [5] for more details on this command.

To illustrate the differences between `Solve` and `Reduce` commands as described above, we consider the following trivial example of finding the solution to the general quadratic equation using *both* commands:

In[1]:=`Solve[{ a*x`^{2}`+b*x+c==0 },{ x }]`

Out[1]=$\left\{ \left\{ x \to \frac{-b-\sqrt{b^2-4ac}}{2a} \right\}, \left\{ x \to \frac{-b+\sqrt{b^2-4ac}}{2a} \right\} \right\}$

In[2]:=`Reduce[{ a*x`^{2}`+b*x+c==0 },{ x }]`

Out[2]=$x = \frac{-b-\sqrt{b^2-4ac}}{2a}$ `&&` $a \neq 0$ `||` $x = \frac{-b+\sqrt{b^2-4ac}}{2a}$ `&&` $a \neq 0$
$a = 0$ `&&` $b = 0$ `&&` $c = 0$ `||` $a = 0$ `&&` $x = -\left(\frac{c}{b}\right)$ `&&` $b \neq 0$

We first note that `&&` represents the "and" operator and `||` represents the "or" operator as in C programming. Then observe that the `Solve` command solves the equation with respect to x and disregards any restrictions on the other terms, e.g., $a \neq 0$ and the `Reduce` command considers all possible scenarios, e.g., the linear solution when $a = 0$.

In using the same procedure to solve (3.50) via the `Reduce` command as outlined above for the `Solve` command, we again obtain the requirement $a_1 = h_2$ in each solution set. As an example, we display the following outputs:

In[1]:=   `Reduce[{ E1==0, E2==0, E3==0, E4==0 },{ a`$_1$`,a`$_2$`,a`$_3$`,h`$_2$` }]`
   Out[1]=$h_2 == a_1$ `&&` $a_1 \neq 0$
In[2]:=   `Reduce[{ E1==0, E2==0, E3==0, E4==0 },{ a`$_2$`,a`$_3$`,h`$_2$` }]`
   Out[2]=$h_2 == a_1$ `&&` $a_1 \neq 0$
In[3]:=   `Reduce[{ E1==0, E2==0, E3==0, E4==0 },{ a`$_1$`,a`$_2$`,a`$_3$` }]`
   Out[3]=$h_2 == a_1$ `&&` $a_1 \neq 0$

We also mention that we additionally omitted the factor $(a_1 - h_2)$ in each of the equations in the system (3.50) and utilized Mathematica® to carry out the same procedure as outlined above using both commands. Once again, these experiments led to the solution set $\{a_1 = h_2\}$. Lastly, the procedure described above was also used to gain insights on the solutions to the other simultaneous systems in Sect. 3.7.

3.9   Conclusion and Future Considerations

To recap, the motivation for conducting this research was threefold. First of all, what has been established herein functions as a template for potentially analyzing other characterization problems. Furthermore, the method utilized in this chapter is elementary as only knowledge of a generating function is needed to conduct the analysis. The only potential shortcoming in this procedure is the prospective complexity of the coefficient comparisons and the solvability of simultaneous system(s) that evolve. However, with modern-day computer algebra systems, like Mathematica®, these spanners can be overcome, as we have shown here. Thus, our secondary motivation was to show how computer algebra aids in the establishment of rigorous results in orthogonal polynomials and special functions.

Third of all, in essence, we have analyzed a special case of the Sheffer *B-Type 1* class and anticipate motivating future researchers to consider further studying more general cases of this class, higher classes, *and* similar characterization problems. Since the theoretical and practical importance of the *B-Type 0* orthogonal polynomial sequences discussed in [4] are so immense, it is quite natural to analyze higher-order Sheffer sequences (and other sequences) in an attempt to determine if novel orthogonal polynomials exist with the utility of the *B-Type 0* classes— Problem 3 at the end of this section states a conjecture related to this.

In order to discuss some future research problems that stem directly from the analysis of this chapter, we point out the following. Given a polynomial sequence $\{P_n(x)\}_{n=0}^{\infty}$ defined by a generating function of the form in Definition 3.4, one can potentially apply the methodology outlined in Sect. 3.3 to obtain necessary conditions on the respective recursion coefficients. Namely, one can develop analogues of Theorems 3.1 and 3.2. From that point, one of two situations can transpire. One, the recursion coefficients are of an unrecognizable form and a novel analysis would have to be completed, possibly like the one outlined in Sects. 3.4–3.7, that would determine the existence/nonexistence of orthogonal sets. Two, the recursion coefficients are in fact recognizable, i.e., they can be manipulated and parameters selected to match known orthogonal sets. In any case, once necessary conditions for the recursion coefficients are established, one would need to show that such conditions are also sufficient and thus, achieve a complete characterization.

In our present work, our method essentially "proves the negative," that is, that orthogonal sets do not exist. As we discussed above, this method can also be used to "prove the positive" as well. Therefore, one can potentially apply our current method to various other polynomials sequences whose generating function is known. In particular, perhaps the most immediate problem of this nature is listed below.

Problem 1. Apply the method of this chapter to the Sheffer *B-Type 0* class:

$$A(t)e^{xH(t)} = \sum_{n=0}^{\infty} S_n(x)t^n$$

with

$$A(t) := \sum_{n=0}^{\infty} a_n t^n, \quad a_0 = 1 \quad \text{and} \quad H(t) := \sum_{n=1}^{\infty} h_n t^n, \quad h_1 = 1$$

with rigorous details. That is, develop analogues of the results in Sect. 3.3 to obtain necessary conditions for the *B-Type 0* recursion coefficients (and then show that these conditions are also sufficient), leading to the most basic complete characterization of the *B-Type 0* class.

This will yield the most basic characterization of the *B-Type 0* class, since only the generating function above (and a three-term recurrence relation) will be needed to obtain the orthogonal sets.

We also have the following:

Problem 2. Apply the methodology of this chapter to the Sheffer *B-Type 2* class:

$$A(t)\exp\left[xH_1(t) + x^2 H_2(t) + x^3 H_3(t)\right] = \sum_{n=0}^{\infty} P_n(x)t^n,$$

$$\text{with } H_i(t) = h_{i,i}t^i + h_{i,i+1}t^{i+1} + \cdots, \quad h_{1,1} \neq 0, \quad i = 1,2,3.$$

This is a perfect supplementation of this chapter. Also, to achieve results regarding a special case of this class, simplifying assumptions can be drawn that are analogous to the ones presented in this chapter. Of course, we can also analyze even higher-order B-Type classes using our current approach; however, Problem 2 will be the next most tractable endeavor.

Lastly, we conclude with the following most general consideration.

Problem 3. Prove the following conjecture:

Conjecture 3.6. The Sheffer *B-Type k* generating function

$$A(t)\exp\left[xH_1(t) + \cdots + x^{k+1}H_{k+1}(t)\right] = \sum_{n=0}^{\infty} P_n(x)t^n,$$

$$\text{with } H_i(t) = h_{i,i}t^i + h_{i,i+1}t^{i+1} + \cdots, \quad h_{1,1} \neq 0, \quad i = 1,2,\ldots,k+1$$

yields orthogonal sets $\{P_n(x)\}_{n=0}^{\infty}$ if and only if $k = 0$.

As was done in this chapter, one would first need to derive expressions for the general Sheffer *B-Type k* recursion coefficients. This of course can be accomplished by using the above generating function and a three-term recurrence relation. From there, one would need to show that these general recursion coefficients must reduce to the *B-Type 0* recursion coefficients. To solve this problem, the analysis in [1] will be beneficial.

References

1. W.A. Al-Salam, *On a Characterization of Meixner's Polynomials*, Quart. J. Math. Oxford, 17(1966), 7–10.
2. G. Gasper, *Using symbolic computer algebra systems to derive formulas involving orthogonal polynomials and other special functions*, in: Nevai, P. (Ed.), Orthogonal Polynomials: Theory and Practice. Kluwer Academic Publishers, Dordrecht, pp. 163–179, (1990).
3. R. Koekoek and R.F. Swarttouw, *The Askey-scheme of hypergeometric orthogonal polynomials and its q-analogue*, Reports of the Faculty of Technical Mathematics and Information, No. 98–17, Delft University of Technology, (1998). http://aw.twi.tudelft.nl/~koekoek/askey/index.html
4. I.M. Sheffer, *Some properties of polynomial sets of type zero*, Duke Math J., 5(1939), 590–622.
5. S. Wolfram, *The Mathematica Book, fifth ed.*, Wolfram Media Inc., Champaign, IL; Cambridge University Press, Cambridge, 2003.
6. S. Wolfram, *Mathematica® 8*, Wolfram Research Inc., 2010.